✦ THE BARNES & NOBLE LIBRARY OF ESSENTIAL READING ✦

TIMAEUS AND CRITIAS

PLATO

TRANSLATED BY BENJAMIN JOWETT

INTRODUCTION BY ODYSSEUS MAKRIDIS

BARNES & NOBLE
NEW YORK

THE BARNES & NOBLE
LIBRARY OF ESSENTIAL READING

Introduction and Suggested Reading
© 2007 by Barnes & Noble, Inc.

Originally published in ca. 355 BC

This 2007 edition published by Barnes & Noble, Inc.

ISBN-13: 978-0-7607-8085-5
ISBN-10: 0-7607-8085-4

Printed and bound in the United States of America

1 3 5 7 9 10 8 6 4 2

CONTENTS

INTRODUCTION

PLATO'S AMBITIOUS DIALOGUE *TIMAEUS* AND THE UNFINISHED *Critias* were meant to be part of a trilogy which, along with the projected *Hermocrates*, would outline a proper and sufficiently detailed natural philosophy and cosmology. The *Timaeus* is Plato's spirited response to the cosmogony and physics of the "atheist" Atomist philosophers Leucippus and Democritus. The *Critias*, which is incomplete, presents what might be a famous Platonic fiction: the story of Atlantis, recounted as a moral metaphor for the cycles of human history. In Plato's philosophy, history and nature are both governed by the order which Reason imposes on an initially chaotic and recalcitrant material universe. Both natural philosophy and philosophic history are, in this view, imbued with rational meaning; the serious reader is expected to gain a proper understanding of moral values in addition to grasping the mechanisms of the material universe and human history. Conversely, failure to study philosophy properly is dangerous for morality, according to Plato.

Following the account of natural philosophy and moral history, the third dialogue *Hermocrates*, which was never written, would probably provide a capstone by laying out principles of political legislation. It is anyone's guess why the *Critias* breaks off unfinished and why the *Hermocrates* was never composed. One explanation is that Plato might have died before he could complete the work. This would mean that this ill-fated trilogy was the last work

Plato ever undertook before he passed on, in 347 BCE at the advanced age of eighty. It has also been suggested, however, that Plato gave up on the project of a trilogy and, instead, pursued a grand synthesis of natural theology and political legislation in his indisputably later work, the *Laws*. Be that as it may, the general impression is that the *Timaeus* is a work composed late in Plato's life. This has been disputed mainly on the ground that certain philosophical problems are presumably picked up and resolved in another dialogue, the *Parmenides*.[1] An erudite response was later offered to this conjecture.[2] It is worth pointing out, on the testimony of late antiquity, that Plato had the habit of revisiting and revising earlier dialogues throughout his life, which makes many a question about dating Platonic dialogues moot.

It has been said of Plato (427–347 BCE) that the history of Western philosophy can be recast as a series of footnotes to his work. This is even more impressive when we take into account that the solutions Plato gives to perennial problems of philosophy are rather outlandish. Even in our times, in the aftermath of Wittgenstein's revolutionary impact on philosophical method, Plato's insights demand serious attention and study by the dedicated student of Western thought. Plato was born into an aristocratic family of Athens and was destined for an ambitious political career. He had been preparing himself through the study of rhetoric and by cultivating his athletic prowess, since good looks counted a great deal in Athens. And then he met Socrates, whom he followed as an obscure and reticent youngster. The rest is history. Plato burned the tokens of his fledgling poetic career and developed a healthy disdain for political oratory. He dedicated the rest of his life to pursuing the life of philosophic investigation, which Socrates taught him to see as the most blessed and highest peak of human achievement. A sworn enemy of the materialist and relativistic philosophers of his time, Plato transcended his early debt to Socratic views and developed his own transcendentalist mystical theory of Forms. In the domain of political thought, Plato is identified with the aristocratic school, having penned a utopia in which the most gifted human natures, who are of course philosophers, are the rulers. Plato's influence survived into the neoplatonic schools of Roman times and exercised decisive influence on the formative years of Christian dogma, which even owes its claims about immortality of the soul to a teaching Plato had inherited from cultic mysteries and, ultimately, from Egypt.

The *Timaeus* is Plato's story of intelligent design. Vastly influential in the history of Western thought until the advent of the modern scientific era, it raises and speculates on philosophical questions about the nature of the universe, the place of humanity in the cosmos, the methods of scientific investigation, and the ultimate purposes of existence. The questions Plato tries to answer in the *Timaeus* are still with us. Why is there a universe rather than mere chaos? How should we explain the presence of life-sustaining processes in a silent and presumably chance-driven universe? Why are the measurable constants of the universe, from gravity and electron charge to the electromagnetic constant and the speed of light, what they are, considering that any deviation would have resulted in a universe that cannot sustain life? How should we account for the apparent goal-oriented behavior of organisms? Can we dispense altogether with teleological explanations in science? What if *assuming* such explanations helps us in our studies? Is there a natural foundation for moral claims? What social consequences are likely to follow from widespread belief in the accidental and desultory nature of material processes? These are the questions encompassed by Plato's ambitious agenda and, regardless of the responses, the challenge is as poignant today as it was in Plato's time.

The action of the *Timaeus* takes place in Athens, in late August, on the day of the greatest panathenean, holiday of the city honoring Athen's patron goddess, Athena. It is supposed to be happening on the day following the action of the Platonic dialogue *Republic*, a brief summary of whose discussion is given. The participants include: Socrates, who in this arguably late dialogue is relegated to a subordinate position; Timaeus of Locri, an adherent of the

Pythagorean school of philosophy, of whom nothing is known and who may well be Plato's invention; Hermocrates, possibly the historic leader who led Syracuse into victory against Athenian forces in the Peloponnesian war; and Critias, who was Plato's great-grandfather. It is mysteriously claimed that an unnamed person, who had been present on the previous day, is absent today.

The *Timaeus* has exercised a pervasive influence on Western thought. This was the only Platonic dialogue known to the Middle Ages, in the form of a Latin translation by Cicero, which is now for the most part lost. Since Plato had presented the *Timaeus* as a response to the Atomists, the Christian Fathers who held sway throughout the Middle Ages found in Plato's work an assault on atheistic claims. Nevertheless, there are irreconcilable differences between monotheistic religious dogma and Plato's cosmology, the most notable being that Plato's divine Craftsman is not omnipotent and the material of the universe is eternal and ungenerated. However, the differences were hushed over by the Christian commentators.

Unsympathetic critics might charge that Plato's influence throughout the Middle Ages retarded the onset of modern science. If only teleological explanations are accepted and all other explanations are rejected, the open-ended, unprejudiced, systematic query into natural causes cannot be pursued. Indeed, the *Timaeus* denounces experiment, which it explains away as a failure to accept the limits of human understanding. On the other hand, the *Timaeus* is not a doctrinaire tract but an instigation to deeper thinking about natural mechanisms and the meaning of human existence, which is guided by the same kind of inquisitive drive that animates not only mystics but also famous scientists. Modern science has proceeded apace in our times, but cannot avoid giving merely peremptory rejoinders to deeper human concerns. It is those concerns that ensure the enduring popularity and relevance of the *Timaeus*.

An ancient controversy, still very much alive, concerns the proper method for reading this seminal Platonic dialogue. Aristotle suggested that a great deal of what is in the *Timaeus* ought to be read metaphorically rather than literally. The camp of the literalists has its own adherents, though, counting such names as the Hellenistic thinker Plutarch among its ranks. In modern times, the metaphorical interpretation has become the preferred approach but it is not so easy to discern Plato's own intentions. A lot depends on how we are to understand Plato's repeated declaration that the account of the dialogue is a "likely account." The word translated as "likely" might have emphatic rather than deflating connotations in the original text, meaning "plausible" or "strongly likely" instead of "merely likely."

Plato's intention is to present an alternative, perhaps the true alternative, to the Atomists. In so doing, Plato is bent on showing how the structured universe itself is the result of Reason's persuasion of blind Necessity. Even if read as a vast metaphor, the dialogue is certainly supposed to elicit the conviction that the best and healthiest state of affairs comes about when and only when we use our intelligence to rule over an intransigent material body. This is what the Demiurge or Craftsman of the dialogue did himself to fashion the ordered and beautiful universe, the *cosmos*, out of primeval chaos.

The Craftsman has to work with chaotic matter much as a carpenter works with an already given material of wood; he cannot create the right kind of wood out of nothing to fit his fancy or serve his purposes. Plato's Craftsman is like a carpenter who applies a rational plan and reaches ingenious compromises with the material, turning the material's own properties to good advantage and sometimes, but not always, managing to bypass its resistance. Nevertheless, many of the undesirable attributes of the material will forever continue to plague the product. Accordingly, our bodily nature reminds us of the insurmountable intransigence of matter, over which even the Craftsman has been unable to prevail completely. Still, the Craftsman's work is the best possible state of affair under the existing material constraints.

The key to the *Timaeus* is what is known as teleology. According to this view, every natural entity has a rational purpose; if normal and healthy, each entity functions to promote its naturally assigned purposes. It follows from this that every satisfactory explanation of biophysiological mechanisms and human action is an explanation in terms of natural purposes or goals. Whatever falls short of the proper natural purposes is to be denounced as defective.

Plato considers the universe to be a living, moving animal (37). The fashioning of perishable things, including the mortal part of human beings, the Craftsman relegated to lesser divinities. References to those deities are not consistent throughout the dialogue and should not be pressed too far.

The Craftsman has an eternal pattern or model after which he fashions the universe. We are reminded here of Plato's theory of Ideas or Forms, with which we are familiar from other dialogues and which is assumed as known to the audience of the *Timaeus*. A Platonic Form is a self-subsisting, eternal, immobile, and unchanging entity that does not exist in natural space or time but in a transcendental realm, aloof from the flux of becoming. It is to a model of this nature that the Craftsman looks up, as it were, in order to fashion the universe, just as an architect, to use yet another metaphor, would apply a blueprint to an initial scattered collection of bricks and mortar.

Two metaphors can be applied to what the Craftsman does: He fashions the discordant matter as one would mold marble to make statues; or, he makes the chaotic universe somehow begin to reflect, appropriately, the corresponding originals or Forms in the way a mirror or surface of a lake can be compelled to reflect images of actual things. The former way of thinking is preferred in the *Timaeus* although traces of the latter can also be found.

Plato claims that there is only one universe, or possible world, and that our universe is the best one that is possible under the persistent constraints of material necessity. Plato tries to prove this as follows: Since the model itself is the best, one of the model's properties must be uniqueness. The copy of the model, the existing universe, must similarly have this property. Therefore, the universe must be not simply one of the better ones but the best possible even if it cannot be, like its original, the best without qualification. It also follows from this that only one structured universe, the best, can be in existence, against the claims of the Atomists who spoke of an infinity of emerging and dissolving local universes. It has been pointed out, however, that the Craftsman would have to be "mad" if he were to try and replicate the property of uniqueness.[3] Certain properties of the Forms, like those that pertain to their privileged status of immutability and timelessness, are proper to the Forms only; they cannot be replicated. The property of uniqueness of the model is one of those properties and the Craftsman should know better than to attempt to copy this property.

The motive of the Craftsman in performing so beneficial an operation is explained. Plato does not refer to a notion of over-flowing charitable love, as later Christian thinkers were to do. Instead, Plato points out that the Craftsman is free of envy, as all superior and self-sufficient creatures are; being above envy, the Craftsman desires that everything should come as close as possible to being like himself.

The Craftsman's operation is described as an act of persuasion of Necessity by superior Reason. Plato exhorts us not to be discouraged on account of the negative tug exercised by matter and to always seek the traces of purposeful, rational, divine action in everything.

The structure of the dialogue reflects the adage of the persuasion of Necessity by Reason. The first part of the dialogue, following a prologue, discusses those things that have been fashioned by Reason. The second part covers the things that come about as a result of the inexorable operation of Necessity. The culmination of the dialogue centers on the monumental work that flows from the cooperation of Reason and Necessity. The conclusion is brief and takes the form of a recapitulation of the main point and exhortation, which can almost be memorized and retained even when memory of the specific arguments has faded.

A number of interesting philosophical problems surround Plato's claim that an intelligent Craftsman, the *Demiurge*, fashioned our universe out of primordial chaos. How was the initial blind and disordered state susceptible of, what Plato calls, rational persuasion? Were there natural laws already at work before the Craftsman's intervention? But how can there be natural laws in a presumably chaotic and disordered medley which has not yet been "persuaded"? On the other hand, however, how can matter be amenable to any organization if it does not demonstrate any regularities in the chaotic state? At a minimum, the properties of the materials used by the Craftsman must be reliable, which means that those properties must be systematically recurring in the original chaotic situation. But, then, how can Plato claim that there was absolute chaos in the original state? •

There is also a problem involving the World-Soul. As mentioned above, Plato sees the universe as a living entity. He speaks of a World-Soul, to which the individual souls are kindred. Is the World-Soul fully rational? Was it rational even before the divine Craftsman intervened? For Plato, rational operations are actually cyclical motions that happen on the same spot. This kind of motion is, for Plato, self-subsistent and self-enabling. Reason and soul do not depend on things outside of themselves and, as such, they are on-the-spot cyclical or rotatory motions. As self-moving, soul is the ultimate source of motion; soul animates and moves everything else. Cyclical motion also stands for Sameness whereas other kinds of motion stand for Difference. Rectilinear motions characterize Difference and are symptomatic of imperfect, alterable, ever-in-flux, and unstable situations. Contemplation involves Sameness but perception depends on Difference. To be sure, the persistence of rectilinear motions in the universe even after the Craftsman has fashioned it as an ordered entity indicates that Reason cannot succeed completely in its persuasion of material Necessity.

Here is the problem involving the World-Soul: Did the chaotic world, before the fashioning of the universe, have a World-Soul? Or did it not? Both answers are unsatisfactory. If there was no

World-Soul, then how was any motion possible? We should remember here that the soul is the origin of all motion. If, on the other hand, the chaotic world did possess a soul of sorts, this soul could not possibly be rational or fully rational. But does this mean that an irrational World-Soul can and has existed? These problems have occupied the attention of Plato's students since the days of his Academy in ancient Athens.

Aristotle himself commented on the difficulties of the *Timaeus*. Aristotle accused Plato of seeking naturalistic explanations for the soul itself. The reason Plato cannot exclude the soul even from the chaotic collection of particles is that Plato takes the soul to be the origin of motion. This, however, means that the difference between Plato and his enemies, the Atomists, shrinks since the Atomists also admit a self-sustaining motion of particles in this universe. This is what bothered Aristotle. Moreover, if the soul is a cause of motion, as Plato argues sometimes, then how can the rational soul be generated, as it is in the *Timaeus*? Things that do not owe their motion to anything else do not need anything external to bring them about. (Of course, in our experience, we never come across any such thing).

A related problem concerns Plato's view of time, on which more follows.

We do not need let the difficulty about the soul detain us but it should be pointed out that several attempted solutions have been proffered. One of the oldest patches was ventured by Plutarch who distinguished between an irrational, and preexisting, soul and the rational, created, soul.

After centuries of speculation, the concepts of space and time still remain mysterious today. Einstein's theory has an odd and ill-defined view of reality, according to which something called spacetime exists, from which positions in space and time can be "decomposed" or computed under specified mathematical procedures.

What has happened to traditional or Newtonian space and time is not clear. But Newton's concepts of space and time, although presumably commonsensical, are themselves plagued by many problems as Newton's critics, like the philosopher Leibniz, obliged to point out right after Newton's physics became publicized.

In the *Timaeus*, Plato is dealing with some of the same problems regarding space and time. It is only because the philosophy, and the physics, of space and time are not studied widely that many readers of Plato's dialogue, including otherwise well-educated people, might be tempted to a sense of smug superiority. If one were to return to Plato's cosmology after careful acquaintance with the bibliography of this abstruse subject, one would have to admire Plato's insight and even his anticipation of subsequent debates.

There are roughly two possible metaphysical positions about space and time. Either space and, more curiously, time are some kinds of real things, no matter how mysterious and elusive they remain; or space and time do not exist, in which case we can talk about them as if they existed because and insofar as human intelligence allows us to do so. More ambitiously, one could argue that space and time do not exist and that relations between different objects or between objects and parts of objects suffice for explaining and accounting for the phenomena we locate in space and time. This latter view is known as Relationism. It had been unpopular until Einstein claimed, partly erroneously, that his theory falls under this category. (In Einstein's theory, the mysterious spacetime manifold is presumably a real entity, albeit a very strange one). On the other hand, the commonsensical or Newtonian view, according to which space and time are real things, is known as Substantivalism.

Plato is a substantivalist about space, which he calls *Chora*, the Receptacle. Plato's view of time, on the other hand, is not Substantivalist in the Newtonian way insofar as Plato allows for certain relational elements. Let us turn to the details.

Plato's view of the Receptacle is no more primitive or less problematic than Newton's. A difference is that Plato's Receptacle is the medium on which the true beings of Platonic metaphysics, the Forms, are expressed: the Receptacle is like a mirror on which the copies of the Forms become possible, or like a material which can be molded to resemble the Forms. The transcendental Forms themselves are not in the space of phenomenal experience. It is not clear if we should say that they are in a space of their own or in no space at all.

According to Plato, the things we know through experience are copies of the Forms; therefore, a medium must exist on which copies can come into (a lesser-degree and borrowed) existence. The Space or Receptacle is this medium. Plato's Receptacle is somewhat like Newton's: the Receptacle is eternal and everlasting, which means that the Receptacle has no origin. Plato's Receptacle, like Newton's space, does not have any characteristics of its own. Plato tries to prove this. If we assume that the Receptacle does have characteristics of its own, then, at some point, the Receptacle would inevitably try to express properties contrary to the ones it has (because, keep in mind, the Receptacle is the only medium and it is the medium for the expression of all properties). When that happened, the medium would be unable to express those properties which are contrary to its own. But—and Plato does not try to prove this point—the Receptacle is capable of expressing *every* property. Therefore, it is impossible that the Receptacle has any properties of its own.

The Receptacle inflates Plato's metaphysics: in addition to Platonic Forms and their perceptible copies, we are told now that there is a third and rather nondescript kind of thing: the Receptacle. This third thing is strange in that it is infinite and omnipresent and yet cannot be perceived or experienced directly. This objection has also been raised against Newton's concept of space. If anything, Plato's Receptacle is not passive like Newton's space: Plato thinks that the Receptacle is actually amenable to molding and has an ability to constrict and agitate the matter that is located in it.

Of course, there is no Form or original of the Receptacle. In terms of Plato's metaphysics, this means that the Receptacle is a privileged kind of thing. The objects of our phenomenal experience all have a borrowed existence and only the Forms are true beings or beings in the full sense of the word. And yet, Plato must accept that there is one thing, the Receptacle, that, although not itself a Form, does not have an extraneous principle of existence.

Plato's view of time is famous and has exercised influence on scores of thinkers. For instance, Plato's concept of time had a profound influence on Augustine's well-known meditation on the problem of the beginning of time, about which Einstein remarked that it is still the limit of our speculative understanding of the subject. Plato also anticipates both Newton's and Einstein's views of time in interesting ways. Plato defines time as the measurable projection of eternity. In eternity everything is static; nothing happens. It is also the case that everything that happens in the dynamic time we experience is in some way a projection of that frozen collection of objects or events that exist in "eternity." As a corollary to Einstein's theory of relativity, modern speculation has also come to see time, in a mysterious union with space, as a block-like entity: in the same way New York is there whether we happen to be visiting or not, the past and the future are also present in an eternal and static "now." The projection of this eternal now constitutes the dynamic flow with which we are familiar in our earthly experience.

A problem that has been briefly mentioned already refers to the generation of time. Plato claims that time begins with the Craftsman's work of ordering the universe. But how can time begin? Doesn't the generation of time presuppose that there is already time so that there is a beginning, or a moment when . . . time begins? Augustine was perplexed by this problem. Moreover, Plato's primordial chaos contains matter tossing about haphazardly; even if the whole motion is disorderly, changes still occur in the chaos and it should be meaningful to speak of the chaotic events as happening *in time*. So, even in the chaotic world, events happen before or after other events, which means that already time is ticking.

A solution has been proposed.[4] We could distinguish two concepts of time in Plato's thought. A) The irreversible temporal succession of moments, which is also, characteristically, the Newtonian concept of time, has no beginning and antedates the imposition of order on the primordial chaos. B) A second concept corresponds to what Aristotle took to be the proper concept of time: uniform flow which can be measured with precision (Aristotle, *Metaphysics* 223a33). Time in the second sense is introduced with the Craftsman's work because only after that point does it become possible to measure the flow of temporal succession in a precise and systematic way. Plato thinks that the underlying foundation of time's measurability is the cyclical or "perfect" motion. Such motions are imposed on matter by the Craftsman and make measurement of temporal succession possible.

To explain the composition of material objects, Plato borrows the particles or indivisible atoms from his theoretical adversaries, the Atomists, and the view that four basic elements exist from the Pre-Socratic philosopher Empedocles. He denies that the particles or atoms are fundamental in nature and tries to explain how Empedocles' four elements (fire, air, water, and earth) come to acquire the specific composition and properties they have. According to Plato, the Craftsman imposed regular geometric shapes on the chaotic primordial matter. Plato thinks that certain geometric figures are "perfect," which is not how today's mathematicians think, of course. Plato tried to solve his own Euclidean geometry theorems in the effort to figure out what geometrical figures are needed: the desideratum is that the figures in question must be inter-transformable, so that changes from one element to the other can be explained. (The only exception Plato makes is for earth, which, he thinks, does not change to anything else).

Plato settled for the "perfect" geometric shapes of pyramid (tetrahedron), octahedron, dodecahedron, icosahedron, and cube. The regularity of these shapes consists in that they are the only three-dimensional figures that have all their surfaces congruent and all their angles equal. The ancient Greeks knew that these solid figures can be inscribed, or fit perfectly, within a sphere—a property which Kepler later utilized to great effect in charting the movements of planets. In Plato's theory, the earth is composed of cubes, water of icosahedrons, air of octahedrons, and fire of tetrahedrons. The one remaining regular solid figure, dodecahedron, is supposed by Plato to correspond to the universe as a whole.

The fundamental constitutive triangles for the composition of the elements are obtained from halving an equilateral triangle and a square: they are, respectively, a half-equilateral triangle and a right-angle isosceles triangle. The recombinations which explain the chemical interactions in the universe are, according to Plato, break-downs of the three-dimensional figures to the basic two-dimensional triangles and recombinations of the latter back into the former. Basic properties of material objects are explained by reference to sharp-ness of angles, fineness of edges, movement, and size. Fire and air penetrate everything with their sharp edges and obtuse angles. In addition, larger bodies have larger gaps in their structures and this causes smaller bodies to combine and act to disintegrate the larger bodies. These processes guarantee perpetual motion of particles.

Plato defines pleasure as the sudden restoration of the natural or normal state and pain as the violent or abrupt disruption of the normal state. Additional principles are assumed: Sensation is the result of motion; and, motion is transferable from a material body to a contiguous one.

Corollaries that follow from Plato's view of pain and pleasure are the following: Gradual transitions from and to normal states are imperceptible. If the transitions are facile but not gradual, the changes are perceptible but neither painful nor pleasurable. (This can help Plato explain why sensations like, for instance, sight, are not painful). Sudden restoration of the normal state is intensely pleasurable and abrupt departure from the normal state is intensely painful.

The body has been fashioned by the auxiliary gods for the sake of providing a "vehicle" for the soul. The rational part of the soul is lodged in the head and, for its protection, the skull is provided. The neck is like a fortification, separating the more noble head from the less dignified trunk and lower body. Again, the mid-section is relatively superior to the lower body. The irrational but potentially

cooperative senses, which came into existence after the incorpora-
tion of the soul, are housed in the mid-section. These are emotions
like fear, anger, courage, and hope. They can be subverted, as when,
for instance, hope strives after obtaining gross and hedonistic
experiences or anger is wrongly directed against one's educators.
But these spirited passions can also be conscripted to the service of
reason, as when, for instance, indignation is vented at a palpable
injustice or fear is used to overawe the carnal desires. Indeed, Plato
takes mid-body organs like the liver to serve purposes of restraint
and discipline. Probably on account of the liver's glossy surface, Plato
thinks that this organ operates as a mirror, reflecting the thoughts of
reason, and in this way offering images which might intimidate the
lower or carnal impulses of the lower body. Additionally, the liver
has a bitter-sweet taste, exercising the bitterness to threaten carnality
and the sweetness to encourage self-restraint. Lastly, the liver can
constrict other organs in order to impose discipline.

The stomach is said to be the manger of the carnal desires. The
intestines are elongated in order to serve as a capacious receptacle of
food, in this way slowing down the voracious, and potentially destructive, human appetite for indulgence. Discussion of the reproductive
organs is oddly postponed, to be taken up near the end of the dialogue in a context in which Plato provides his most extreme, and
infamous, misogynist statement—that women came into existence
for the first time only in the second generation, as reincarnations of
evil-doing men.

Even the basest parts, Plato points out, are in the "right path,"
having been fashioned by deities and according to the best model.
The success is vitiated by the ineluctable operation of the characteristics of a material body: Necessity has been "persuaded" to
serve Reason but the truce is one of halting cooperation rather than
a triumph for Reason. An interesting example of a second-best
compromise is this: For the protection of the brain, thickness of
skin is mandated, and has been provided to some extent. But the
brain also needs mobility so that it can receive transmitted

motion, which is needed for keen perception. Human beings would have longer lives and be much better protected from harm if the skull had been impregnable, but then human life would be impoverished since perception would be severely impaired. A compromise between the best possible state of affairs and what is necessary has been reached in making the skull of medium rather than extreme thickness.

Plato's psychology in the *Timaeus*, as in the earlier dialogue *Republic*, violates the principle defended in the early dialogue *Phaedo*, according to which the soul is a simple, uncompounded entity. Now the soul is assumed to have parts, most notably a rational part and two others, the spirited and the lowest one which contains the carnal appetites and desires. The *Timaeus* reiterates this tripartition of the soul and adds a new twist: only the rational part of the soul is immortal whereas the other two perish when soul is severed from body in death.

The unfinished dialogue *Critias*, which follows the *Timaeus*, highlights Plato's intention to provide a moral lesson to his audience or readers. The famous Atlantis story, probably invented by Plato, records the vicissitudes of a powerful island nation somewhere in the Atlantic, whose inhabitants lapsed into insolence and decadence. In the Atlantis story, Plato casts Athens in the role of righteous punisher of the Atlantis. Plato's point, however, is not to celebrate Athenian imperialism, which would only repeat the Atlantis predicament of arrogant power, but to remind his readers that there was a time when the Athenians were leading a life of justice and proper measure. The story of how the structured universe itself came into existence adds credence to this urgent Platonic message: we should live our lives always following the guidance of reason and prevailing over the indulgent and destructive tendencies of our material bodies.

Odysseus Makridis received his Ph.D. from Brandeis University. He is Assistant Professor of Philosophy at Fairleigh Dickinson University, in Madison, New Jersey.

Editor's Note

The numeric marginalia found within this text are Stephanus numbering. In 1578, Henri Estienne, commonly known by the Latin name Stephanus, assembled an edition of Plato's dialogues that became the standard for scholars. Due to translation variances and reordering of the text, the numeric marginalia of this Barnes & Noble edition, like many modern editions, are not necessarily chronological. The Stephanus numbers in the margin therefore assist in standardizing the way scholars refer to Plato's original, Greek text.

TIMAEUS

Persons of the Dialogue

SOCRATES.   CRITIAS.
TIMAEUS.   HERMOCRATES.

SOC. ONE, TWO, THREE; AND WHERE, MY DEAR TIMAEUS, IS THE
fourth of those who were yesterday our guests and are to be our
entertainers today?

TIMAEUS. He has been taken ill, Socrates, or he certainly would not
have been absent at such a meeting as this.

SOC. Then, if he is not coming, you and the two others must supply
his place.

TIM. Assuredly we will do all that we can; having been handsomely
entertained by you yesterday, we who remain ought gladly to entertain you in return.

SOC. Do you remember how many points there were of which I told
you that we must speak?

TIM. We remember some of them, and you will be able to remind us
of what we may have forgotten: or rather, if we are not troubling
you, will you briefly recapitulate the whole, and then we shall be
more certain?

SOC. I was inquiring yesterday how and of what citizens the best
State would be composed, that was the main purpose of what I
was saying.

TIM. I am sure, Socrates, that your words were very much to
our mind.

Soc. Do you remember the part about the husbandmen and the artisans; and how we began by separating them from the class of defenders of the State?

Tim. Yes.

Soc. And when we had given to each one that single employment and particular art which were suited to his nature, we spoke of those who were intended to be our warriors, and said that they were to be guardians of the city against the attacks of enemies internal as well as external, and to have no other employment; with gentleness they were to judge their subjects, of whom they were by nature friends, but when they came in the way of their enemies in battle they were to be fierce with them.

Steph.
18

Tim. Exactly.

Soc. We said, if I am not mistaken, that the guardians should be gifted with a passionate and also with a philosophical temper, and that this would be the true way of making them gentle to their friends and fierce to their enemies.

Tim. Certainly.

Soc. And what did we say of their education? Were they not to be trained in gymnastic, and music, and all other sorts of knowledge which were proper for them?

Tim. Very true.

Soc. Thus trained, our citizens were not to think of gold and silver, or any other possession as their own private property; they were to be hired troops, receiving pay for keeping guard from those who were protected by them—the pay was to be no more than would suffice for men of simple life; and they were to have their expenses in common, and to live together in the continual practice of virtue, which was to be their sole pursuit.

Tim. That also was said.

Soc. Neither did we forget the women, of whom we said, that their natures should be made as nearly as possible like those of men, and that they should share with them in their military pursuits and in their general way of life.

TIM. That, again, was as you describe.

SOC. And what was said about the procreation of children? That was too singular to be easily forgotten, for the proposal was that all wives and children should be in common; and we devised means that no one should ever be able to know his own child, but that all should imagine themselves to be of one family, and should regard as brothers and sisters those who were within a certain limit of age; and those who were of an elder generation they were to regard as parents and grandparents, and those who were of a younger generation as children and grandchildren.

TIM. Yes, indeed, there is no difficulty in remembering that.

SOC. And do you also remember how, with a view of having as far as we could the best breed, we said that the chief magistrates, male and female, were to contrive secretly, by the use of certain lots, so to arrange the nuptial meeting, that the bad of either sex and the good of either sex should pair with their like, and there was to be no quarrelling on this account, for they were to imagine that the union was a mere chance, and was to be attributed to the lot?

TIM. I remember.

SOC. And you remember how we said that the children of the good parents were to be educated, and the children of bad parents secretly dispersed among the other citizens, and when they began to grow up the rulers were to be on the lookout, and to bring up from below in their turn those who were worthy, and those among themselves who were unworthy were to take the places of those who came up? 19

TIM. True.

SOC. Then have I now given you a complete summary of our yesterday's discussion? Or is there anything, dear Timaeus, that has been omitted?

TIM. No, Socrates, those were precisely the heads of the discussion.

SOC. Then let me now proceed to tell you my own feeling about the State which we have described. I might compare myself to a person who, on beholding beautiful animals either created by the painter's art, or really alive but at rest, is seized with a desire of beholding them in motion or engaged in some struggle or conflict to which their forms appear suited. This is my feeling about the State which we have described: there are conflicts which all cities undergo, and I should like to hear someone tell of our own city carrying on a struggle against her neighbors, and how she went out to war in a fitting manner, and when at war showed a result answerable to her training and education, both in her modes of action and fashions of speech, when dealing with other cities. Now I, Critias, and Hermocrates, am conscious that I myself should never be able to set forth the city and her citizens in proper terms of praise, and I am not surprised at my own incapacity; to me the wonder is rather that the poets, present as well as past, are no better—not that I mean to depreciate them, but everyone can see that they are a tribe of imitators, and will imitate best and most easily that in which they have been brought up; whereas that which is beyond the range of a man's education can hardly be imitated by him in action, and with still more difficulty in speech. I am aware that the Sophists have plenty of brave words and fair devices, but I am afraid that being only wanderers from one city to another, and having never had homes of their own to manage, they may err in their ideas of philosophers and statesmen, and may fail to know what they do and say in their dealings with mankind on all the various occasions of peace and war. And thus people of your class are the only ones remaining who are fitted by nature and education to take part at once in politics and philosophy. Here is Timaeus, of Locris in Italy, a city which has excellent laws, and who is himself in wealth and rank the equal of any of his fellow citizens; he has held the most important and honorable offices in his own State, and, as I believe, has scaled the heights of philosophy; and here is Critias, whom every Athenian knows to be well acquainted with the things of which we are speaking; and

20

as to Hermocrates, I am assured by many witnesses that he is by nature and education well suited to philosophical inquiries. And therefore yesterday when I saw that you wanted me to discuss the formation of the State I readily complied, being very well aware, that, if you only would, none were better qualified to carry the discussion further, and that when you had engaged our city in a suitable war, you of all men living could best exhibit her playing her part in that situation. Having now completed my task, I in return impose this other task upon you. There was an agreement that you were to entertain me as I have entertained you. Here am I in festive array, and no man can be more ready for the promised banquet.

HER. And we too, Socrates, as Timaeus says, will do our utmost; there would be no excuse for our refusal to comply. As soon as we arrived yesterday at the guest-chamber of Critias, with whom we are staying, or rather on our way thither, we talked the subject over, and he told us an ancient tradition, which I wish, Critias, that you would repeat to Socrates, and then he will be able to judge whether it fulfills his requirements.

CRIT. That I will, if Timaeus, who is our partner, approves.

TIM. I approve.

CRIT. Then listen, Socrates, to a strange tale which is, however, certainly true, as Solon, who was the wisest of the seven sages, declared. He was a relative and a great friend of my great-grandfather, Dropidas, as he himself says in several of his poems; and Dropidas told Critias, my grandfather, who remembered and told us: That there were of old great and marvelous actions of the Athenians, which have passed into oblivion through time and the destruction of the human race, and one in particular, which was the greatest of them all, the recital of which will be a suitable testimo- 21 ny of our gratitude to you, and also a hymn of praise true and worthy of the goddess, which may be sung by us at the festival in her honor.

Soc. Very good. And what is this ancient famous action of which Critias spoke not as a mere legend, but as a veritable action of the Athenian State, which Solon recounted?

Crit. I will tell an old-world story which I heard from an aged man; for Critias was, as he said, at that time nearly ninety years of age, and I was about ten years of age. Now the day was that day of the Apaturia which is called the registration of youth, at which, according to custom, our parents gave prizes for recitations, and the poems of several poets were recited by us boys, and many of us sang the poems of Solon, which were new at the time. One of our tribe, either because this was his real opinion, or because he thought that he would please Critias, said that in his judgment Solon was not only the wisest of men, but also the noblest of poets. The old man, as I very well remember, brightened up at this and said, smiling: "Yes, Amynander, if Solon had only, like other poets, made poetry the business of his life, and had completed the tale which he brought with him from Egypt, and had not been compelled, by reason of the factions and troubles which he found stirring in this country when he came home, to attend to other matters, in my opinion he would have been as famous as Homer or Hesiod, or any poet."

"And what was the poem about, Critias?" said the person who addressed him.

"About the greatest action which the Athenians ever did, and which ought to have been the most famous, but which, through the lapse of time and the destruction of the actors, has not come down to us."

"Tell us," said the other, "the whole story, and how and from whom Solon heard this veritable tradition."

He replied: "At the head of the Egyptian Delta, where the river Nile divides, there is a certain district which is called the district of Sais, and the great city of the district is also called Sais, and is the city from which Amasis the king was sprung. And the citizens have a deity who is their foundress; she is called in the Egyptian tongue Neith, and is asserted by them to be the same whom the Hellenes called Athene. Now the citizens of this city are great lovers of the Athenians, and say that they are in some way related to them. Thither came Solon, who was received by them with great honor; and he asked the priests, who were most skillful in such matters, about antiquity, and made the discovery that neither he nor any other Hellene knew anything worth mentioning about the times of old. On one occasion, when he was drawing them on to speak of antiquity, he began to tell about the most ancient things in our part of the world—about Phoroneus, who is called 'the first,' and about Niobe; and after the Deluge, to tell of the lives of Deucalion and Pyrrha; and he traced the genealogy of their descendants, and attempted to reckon how many years old were the events of which he was speaking, and to give the dates. Thereupon, one of the priests, who was of a very great age, said: 'O Solon, Solon, you Hellenes are but children, and there is never an old man who is an Hellene.' Solon hearing this, said, 'What do you mean?' 'I mean to say,' he replied, 'that in mind you are all young; there is no old opinion handed down among you by ancient tradition; nor any science which is hoary with age. And I will tell you the reason of this. There have been, and will be again, many destructions of mankind arising out of many causes; the greatest have been brought about by the agencies of fire and water, and other lesser ones by innumerable other causes. There is a story, which even you have preserved, that once upon a time Phaëthon, the son of Helios, having yoked the steeds in his father's chariot, because he was not able to drive them in the path of his father, burnt up all that was upon the earth, and was himself destroyed by a thunderbolt. Now, this has the form of a myth, but really signifies a declination of the bodies moving around the earth and in the heavens, and a great conflagration of things upon the earth recurring at long intervals of time; when this happens, those who live upon the mountains and in dry and lofty places are more liable to destructions than those who dwell by rivers or on the

22

seashore. And from this calamity the Nile, who is our never-failing savior, saves and delivers us. When, on the other hand, the gods purge the earth with a deluge of water, among you, herdsmen and shepherds on the mountains are the survivors, whereas those of you who live in cities are carried by the rivers into the sea. But in this country, neither at that time nor at any other, does the water come from above on the fields, having always a tendency to come up from below, for which reason the things preserved here are said to be the oldest. The fact is, that wherever the extremity of winter frost or of summer sun does not prevent, the human race is always increasing at times, and at other times diminishing in numbers.

23 And whatever happened either in your country or in ours, or in any other region of which we are informed—if any action which is noble or great or in any other way remarkable has taken place, all that has been written down of old, and is preserved in our temples; whereas you and other nations are just being provided with letters and the other things which States require; and then, at the usual period, the stream from heaven descends like a pestilence, and leaves only those of you who are destitute of letters and education; and thus you have to begin all over again as children, and know nothing of what happened in ancient times, either among us or among yourselves. As for those genealogies of yours which you have recounted to us, Solon, they are no better than the tales of children; for in the first place you remember one deluge only, whereas there were many of them; and in the next place, you do not know that there dwelt in your land the fairest and noblest race of men which ever lived, of whom you and your whole city are but a seed or remnant. And this was unknown to you, because for many generations the survivors of that destruction died and made no sign. For there was a time, Solon, before the great deluge of all, when the city which now is Athens, was first in war and was preëminent for the excellence of her laws, and is said to have performed the noblest deeds and to have had the fairest constitution of any of which tradition tells, under the face of heaven.' Solon marveled at

this, and earnestly requested the priest to inform him exactly and in order about these former citizens. 'You are welcome to hear about them, Solon,' said the priest, 'both for your own sake and for that of the city, and above all, for the sake of the goddess who is the common patron and protector and educator of both our cities. She founded your city a thousand years before ours, receiving from the Earth and Hephaestus the seed of your race, and then she founded ours, the constitution of which is set down in our sacred registers as 8,000 years old. As touching the citizens of 9,000 years ago, I will briefly inform you of their laws and of the noblest of their actions; and the exact particulars of the whole we will hereafter go through at our leisure in the sacred registers themselves. If you compare these very laws with your own you will find that many of ours are the counterpart of yours as they were in the olden time. In the first place, there is the caste of priests, which is separated from all the others; next there are the artificers, who exercise their several crafts by themselves and without admixture of any other; and also there is the class of shepherds and that of hunters,[1] as well as that of husbandmen; and you will observe, too, that the warriors in Egypt are separated from all the other classes, and are commanded by the law only to engage in war; moreover, the weapons with which they are equipped are shields and spears, and this the goddess taught first among you, and then in Asiatic countries, and we among the Asiatics first adopted. Then as to wisdom, do you observe what care the law took from the very first, searching out and comprehending the whole order of things down to prophecy and medicine (the latter with a view to health); and out of these divine elements drawing what was needful for human life, and adding every sort of knowledge which was connected with them. All this order and arrangement the goddess first imparted to you when establishing your city; and she chose the spot of earth in which you were born, because she saw that the happy temperament of the seasons in that land would produce the wisest of men. Wherefore the

24

goddess who was a lover both of war and of wisdom, selected and first of all settled that spot which was the most likely to produce men likest herself. And there you dwelt, having such laws as these and still better ones, and excelled all mankind in all virtue as became the children and disciples of the gods.

"'Many great and wonderful deeds are recorded of your State in our histories. But one of them exceeds all the rest in greatness and valor. For these histories tell of a mighty power which was aggressing wantonly against the whole of Europe and Asia, and to which your city put an end. This power came forth out of the Atlantic Ocean, for in those days the Atlantic was navigable; and there was an island situated in front of the straits which you call the columns of Heracles; the island was larger than Libya and Asia put together, and was the way to other islands, and from the islands you might pass through the whole of the opposite continent which surrounded the true ocean; for this sea which is within the Straits of Heracles is only a harbor, having a narrow entrance, but that other is a real sea, and the surrounding land may be most truly called a continent. Now in this island of Atlantis there was a great and wonderful empire which had rule over the whole island and several others, as well as over parts of the continent, and, besides these, they subjected the parts of Libya within the columns of Heracles as far as Egypt, and of Europe as far as Tyrrhenia. The vast power thus gathered into one, endeavored to subdue at one blow our country and yours and the whole of the land which was within the straits; and then, Solon, your country shone forth, in the excellence of her virtue and strength, among all mankind; for she was the first in courage and military skill, and was the leader of the Hellenes. And when the rest fell off from her, being compelled to stand alone, after having undergone the very extremity of danger, she defeated and triumphed over the invaders, and preserved from slavery those who were not yet subjected, and freely liberated all the others who dwell within the limits of Heracles. But afterwards there occurred violent

25

earthquakes and floods; and in a single day and night of rain all your warlike men in a body sank into the earth, and the island of Atlantis in like manner disappeared, and was sunk beneath the sea. And that is the reason why the sea in those parts is impassable and impenetrable, because there is such a quantity of shallow mud in the way; and this was caused by the subsidence of the island.'"

I have told you shortly, Socrates, the tradition which the aged Critias heard from Solon. And when you were speaking yesterday about your city and citizens, this very tale which I am telling you came into my mind, and I could not help remarking how, by some coincidence not to be explained, you agreed in almost every particular with the account of Solon; but I did not like to speak at the moment. For as a long time had elapsed, I had forgotten too much, and I thought that I had better first 26 of all run over the narrative in my own mind, and then I would speak. And for this reason I readily assented to your request yesterday, considering that I was pretty well furnished with a theme such as the audience would approve, and to find this is in all such cases the chief difficulty.

And therefore, as Hermocrates has told you, on my way home yesterday I imparted my recollections to my friends in order to refresh my memory, and during the night I thought about the words and have nearly recovered them all. Truly, as is often said, the lessons which we have learned as children make a wonderful impression on our memories, for I am not sure that I could remember all that I heard yesterday, but I should be much surprised if I forgot any of these things which I have heard very long ago. I listened to the old man telling them, when a child, with great interest at the time; he was very ready to teach me, and I asked him about them a great many times, so that they were branded into my mind in ineffaceable letters. As soon as the day broke I began to repeat them to my companions, that they as well as myself might have a material of discourse. And now, Socrates, I am ready to tell you the whole tale of which this is the introduction. I will give you not only the general heads, but the details exactly as I heard them. And as to the city and citizens, which you yesterday described to us in fiction, let us transfer them to the world of reality; this shall be our city, and we will suppose that the citizens whom you imagined, were our veritable ancestors—the same of whom the priest was telling; they will perfectly agree, and there will be no inconsistency in saying that the citizens of your republic are these ancient Athenians. Let us distribute the discussion amongst us, and all endeavor as far as we can to carry out your instructions. Consider then, Socrates, if this narrative is suited to the purpose, or whether we should seek for some other instead.

Soc. And what other, Critias, can we find that will be better than this which is natural and suitable to the festival of the goddess; and has the very great advantage of being a fact and not a fiction? How or where shall we find others if we abandon this? There are none to be had, and therefore you must tell the tale, and good luck to you; and I in return for my yesterday's discourse will now rest and be a listener.

27

Crit. Let me proceed to explain to you, Socrates, the order in which we have arranged our entertainment. The intention is that Timaeus, who is the most of an astronomer amongst us, and has made a special study of the nature of the universe, should speak first, beginning with the generation of the world and going down to the creation of man; next, I am to receive the citizens of whom he is the imaginary parent, and some of whom will have profited by the excellent education which you have given them; and then, in conformity with the law of Solon, we will bring the heroes of his tale into court and judge them ourselves, as if they were those very Athenians whom the sacred Egyptian record has recovered from oblivion, and we shall thenceforward be entitled to speak of them as Athenians and fellow citizens.

Soc. I see that I shall receive in my turn a perfect and noble feast of reason. And now, Timaeus, you I suppose are to follow, first offering up a prayer to the gods as is customary.

Tim. All men, Socrates, who have any degree of right feeling do this at the beginning of every enterprise great or small—they always call upon the gods. And we, too, who are going to discourse of the nature of the universe, whether created, or uncreated, if we be not altogether out of our wits, must invoke and pray the gods and goddesses that we may say all things in a manner pleasing to them and consistent with ourselves. Let this, then, be our invocation to the gods, to which I add an exhortation to myself that I may set forth this high argument in the manner which will be most intelligible to you, and will most accord with my own intent.

First, if I am not mistaken, we must determine, What is that which always is and has no becoming; and what is that which is always becoming and has never any being? That which is apprehended by reflection and reason always is, and is the same; that, on the other hand, which is conceived by opinion 28 with the help of sensation and without reason, is in a process of becoming and perishing, but never really is. Now everything that becomes or is created must of necessity be created by some cause, for nothing can be created without a cause. That of which the artificer looks always to the same and unchangeable, and of which he works out the form and nature after an unchangeable pattern, must of necessity be made fair and perfect; but that of which the artificer looks to the created only, and fashions after a created pattern, is not fair or perfect. Was the heaven then or the world, whether called by this or any other more appropriate name—the question which I am going to ask has to be asked about the beginning of everything—was the world, I say, always in existence and without beginning? Or created and having a beginning? Created, I reply, being visible and tangible and having a body, and therefore sensible; and all sensible things which are apprehended by opinion and sense are in process of creation and created. Now that which is created must of necessity be created by a cause. But how can we find out the father and maker of all this universe? Or when we have found him how shall we be able to speak of him to all men? And there is still another question to be asked about him, Which of the patterns had the artificer in view when he made the world, the pattern which is unchangeable, or that which is created? If the world be indeed fair and the artificer good, then, as is plain, he must have looked to that which is eternal. But if what I may not venture to say is true, then he looked to the created pattern. Everyone will see that he must have looked to the eternal, for the world is the fairest of creations and he is the best of causes. And being of such a nature the world has been framed by him with a view to that which is apprehended by reason and mind and is unchangeable, and if this be admitted must of necessity be the copy of something. Now that the beginning of everything should be according to nature is a great matter. Let us then assume about the copy and original that the words are akin to the matter which they describe, and that when they relate to the lasting and permanent and intelligible, they ought to be lasting and unfailing, and as far as is in the nature of words irrefutable and immovable, and nothing less than this. But the words which are the expression of the imitation of the eternal things, which is an image only, need only be likely and analogous to the former words. What essence is to generation, that, truth is to belief. If then, Socrates, amid the many opinions about the gods and the generation of the universe, we are not able to give notions that are in every way exact and consistent with one another, do not wonder at that. If only we adduce probabilities as likely as any others, that ought to be enough for us, when we remember that I who am the speaker, and you who are the judges are only mortal men, and we ought to accept the tale which is probable and not inquire further.

29

Soc. Excellent, Timaeus, your words shall be taken as you mean them. We have heard your prelude with the greatest pleasure, and now beg you to proceed to the strain which follows.

Tim. Let me tell you then, why the creator of the world generated and created this universe. He was good, and no goodness can ever have any jealousy of anything. And being free from jealousy, he desired that all things should be as like himself as possible. This is the true beginning of creation and of the world, which we shall do well in receiving on the testimony of wise men: God desired that all things should be good and nothing bad as far as this could be accomplished. Wherefore also finding the whole visible sphere not at rest, but moving in an irregular and disorderly manner, out of disorder he brought order, considering that this was far better than the other. Now he who is the best neither creates nor ever has created anything but the fairest, and reflecting upon the visible works of nature, he found that no unintelligent creature taken as a whole was fairer than the intelligent taken as a whole; and that intelligence could not exist in anything which was devoid of soul. For these reasons he put intelligence in soul, and soul in body, and framed the universe to be the best and fairest work in the order of nature. And therefore using the language of probability, we may say that the world became a living soul and truly rational through the providence of God.

30

This being supposed, let us next proceed to consider the further question, in the likeness of what animal did the Creator make the world? Certainly we cannot suppose that the form was like that of beings which exist in parts only; for nothing can be beautiful that is like any imperfect thing; but we may regard the world as the very likeness of that of which all other animals, both individually and as tribes are portions. For the pattern of the universe contains in itself all intelligible beings, just as this world contains us and all other visible creatures. For the Deity intending to make this world like the fairest and most perfect of intelligible beings, framed one visible animal comprehending within all other animals of a kindred nature. Are we right in saying that there is one heaven, or shall we rather say that they are many and infinite? There is one, if the created heaven is to accord with the pattern. For that which includes all other intelligible creatures cannot have a second or companion;

31

in that case there would be need of another living being which would include those two, and of which they would be parts, and the likeness would be more truly said to resemble not those two, but that other which included them. In order then that the world might be like the perfect animal in unity, he who made the worlds made them not two or infinite in number; but there is and ever will be one only-begotten and created heaven.

Now that which is created is of necessity corporeal, and also visible and tangible. And nothing is visible when there is no fire, or tangible which is not solid, and nothing is solid without earth. Wherefore also, God in the beginning of creation made the body of the universe to consist of fire and earth. But two things cannot be held together without a third; they must have some bond of union. And the fairest bond is that which most completely fuses and is fused into the things that are bound; and proportion is best adapted to effect such a fusion. For whenever in three numbers, whether solids or of any other power, there is a mean, and the mean is to the last term what the first term is to the mean; and again, when the mean is to the first term as the last term is to the mean, then the mean becoming first and last, and the first and last both becoming means, all things will of necessity come to be the same, and being the same, with one another will all be one. If now the universal frame had been created a surface only and had no depth, one mean would have sufficed to bind together itself and the other terms; but now, as the world must be solid, and solid bodies are always compacted not by one mean but by two, God placed water and air in the mean between fire and earth, and made them to have the same proportions as far as was possible (as fire is to air so is air to water, and as air is to water so is water to earth), and thus he bound and put together a visible and palpable heaven. And for these reasons, and out of these elements, which are in number four, the body of the world was created in the harmony of proportion, and therefore having the spirit of

32

friendship; and being at unity with itself, was indissoluble by the hand of any other than the framer.

Now the creation took up the whole of each of the four elements; for the Creator compounded the world out of all the fire and all the water and all the air and all the earth, leaving no part of any of them nor any power of them outside. He intended, in the first place, that the whole animal should be perfect, as far as possible, and that the parts of which he was formed should be perfect; and that he should be one, 33 leaving no remnants out of which another such world might be created: and, also, that he should be free from old age and unaffected by disease. And, considering that hot and cold and other powerful forces which unite bodies are apt to surround and attack them from without when they are unprepared, and by bringing diseases and old age upon them, make them to dissolve and die, for this cause and on these grounds, he fabricated the world whole and of whole elements, perfect and not liable to old age and disease. And he gave to the world the figure which was suitable and also natural. But, to the animal which was to comprehend all animals, that figure was suitable which comprehends within itself all other figures. Wherefore also he made the world in the form of a globe, round as from a lathe, in every direction equally distant from the center to the extremes, the most perfect and the most like itself of all figures; for he considered that the like is infinitely fairer than the unlike. This he finished all round, and made the outside quite smooth for many reasons; in the first place, because eyes would have been of no use to him when there was nothing remaining without him, or which could be seen; and there would have been no use in ears when there was nothing to be heard; nor was there any surrounding atmosphere to be breathed; nor would there have been any use of implements by the help of which he might receive his food or get rid of what he had already digested; for there was nothing which went from him or came to him, seeing that there was nothing beside him. And he himself provided his nutriment to himself through his own decay, and all that he did or suffered was done in himself and by himself, according to art. For the Creator conceived that a being which was self-sufficient would be far more excellent than one that lacked anything; and, as he had no need to take anything or defend himself against anyone, he had no need of hands, and the Creator did not think necessary to furnish him with them when he did not want them: nor had he any need of feet, nor of the whole apparatus of walking; but he assigned to him the motion appropriate to his spherical form, being that of all the seven which is most appropriate to mind and intelligence, and so he made him move in the same manner and on the same spot, going round in a circle turning within himself. All the other six motions he took away from him, and made him incapable of being affected by them. And as this circular movement required no feet, he made the universe without feet or legs.

34

Such was the whole scheme of the eternal God about the god that was to be, to whom he for all these reasons gave a body, smooth, even, and in every direction equidistant from a center, entire and perfect, and formed out of perfect bodies. And in the center he put the soul, which he diffused through the whole, and also spread over all the body round about; and he made one solitary and only heaven a circle moving in a circle, having such excellence as to be able to hold converse with itself, and needing no other friendship or acquaintance. Having these purposes in view he created the world to be a blessed god.

Now God did not make the soul after the body, although we have spoken of them in this order; for when he put them together he would never have allowed that the elder should serve the younger, but this is what we say at random, because we ourselves too are very largely affected by chance. Whereas he made the soul in origin and excellence prior to and older than

the body, to be the ruler and mistress, of whom the body was
to be the subject. And the soul he made out of the following elements and on this manner: he took of the ^{35}
unchangeable and indivisible essence, and also of the divisible
and corporeal which is generated, and he made a third sort
of intermediate essence out of them both, partaking of the
nature of the same and of the other, and thus he compounded
a nature which was in a mean between the indivisible and the
divisible and corporeal. These three elements he took and
mingled them all in one form, compressing the reluctant
and unsociable nature of the other into the same. And when
he had mixed them with the essence and out of all the three
made one, he again divided this whole into as many portions
as was fitting, each of them containing an admixture of the
same and of the other and of the essence. And he began to
divide on this wise: first of all, he took away one part of the
whole and then he separated

[1, 2, 3, 4, 9, 8, 27,]

a second part which was double the first, and then he took away a
third part which was half as much again as the second and three
times as much again as the first, and then he took a fourth
part which was twice as much as the second, and a fifth part
which was three times as much as the third, and a sixth part which
was eight times as much as the first, and a seventh part which was
twenty-seven times the first. After this he filled up the ^{36}
double and triple intervals [i.e., 1, 2, 4, 8, and 1, 3, 9, 27],
cutting off portions from the whole and placing them between
the intervals, so that in each interval there were two kinds
of means—

$$[\overline{1}, \tfrac{4}{3}, \tfrac{3}{2}, \overline{2}, \tfrac{8}{3}, 3, \overline{4}, \tfrac{16}{3}, 6, \overline{8},]$$
$$[\overline{1}, \tfrac{3}{2}, 2, \overline{3}, \tfrac{9}{2}, 6, \overline{9}, \tfrac{27}{2}, 18, \overline{27},]$$

the one exceeding and exceeded by equal parts of the respective extremes [as for example 1, $\frac{4}{3}$, 2, in which the mean $\frac{4}{3}$ is one third more than 1 and one third less than 2], the other being that kind of mean which exceeds and is exceeded by an equal number. Where there are intervals of $\frac{3}{2}$ and of $\frac{4}{3}$ and of $\frac{9}{8}$, made by the connecting terms in the former intervals, he filled up all the intervals of $\frac{4}{3}$ with the intervals $\frac{9}{8}$, leaving a part of each, of which the interval was in the ratio of 256 to 243.[2] And thus the whole mixture out of which he cut these portions was all exhausted by him. This entire compound he divided lengthways into two parts, which he joined to one another at the center like the figure of a)(, and bent them into a circular form, connecting them with themselves and each other at the point opposite to that of contact; and, comprehending them in an uniform motion on the same spot around a center, he made the one the outer and the other the inner circle. Now the motion of the outer circle he called the motion of the same, and the motion of the inner circle the motion of the other. The motion of the same he made to proceed round by the side to the right, and the motion of the other diagonally to the left. And he gave dominion to the motion of the same and the like, for that he left single and undivided; but the inner motion he split into six portions and made seven unequal orbits, having their intervals in ratios of two and three, three of each, and bade their orbits move in a direction opposite to one another: and three [Sun, Mercury, Venus] he made to move with equal swiftness, and the remaining four [Moon, Saturn, Mars, Jupiter] to move with unequal swiftness to the three and to one another, but all in due course.

Now when the Creator had framed the soul according to his will, he formed within the mind the corporeal universe, and brought them together, and united them center to center. The soul, interfused everywhere from the center to the circumference of heaven, of which she is the external envelopment,

herself turning in herself, began a divine beginning of never-
ceasing and rational life enduring throughout all time. The
body of heaven is visible, but the soul invisible, and partakes
of reason and harmony, and being made by the best of
intelligible and everlasting beings, is the best of things   37
created. And as being composed of the nature of the same
and of the other and of the essence, these three, and divided and
bound together in proportion, and revolving in the circle of
herself, the soul, when touching anything which has essence,
whether dispersed in parts or undivided, is stirred throughout
her being to declare the sameness and diversity of things, and
as to what and in what way and how and when individuals are
related or affected, both in the world of generation and in the
world of immutable being. And when reason, which works with
equal truth both in the circle of the other and of the same, in
the sphere of the self-moved in voiceless silence moving, when
reason, I say, is in the neighborhood of sense, and the circle of
the other also moving truly[3] imparts the intimations of sense
to the whole soul, then arise fixed and true opinions and beliefs.
But when reason is in the sphere of the rational, and the circle of
the same moving smoothly indicates this, then intelligence and
knowledge are of necessity perfected. And if anyone affirms that
in which these are found to be other than the soul, he will say
the very opposite of the truth.

When the father and creator saw the image that he had made
of the eternal gods moving and living, he was delighted, and
in his joy determined to make his work still more like the
pattern; and as the pattern was an eternal creature, he sought
to make the universe the same as far as might be. Now the
nature of the intelligible being is eternal, and to bestow eternity on the creature was wholly impossible. But he resolved
to make a moving image of eternity, and as he set in order the
heaven he made this eternal image having a motion according
to number, while eternity rested in unity; and this is what we call

time. For there were no days and nights and months and years before the heaven was created, but when he created the heaven he created them also. All these are the parts of time, and the past and future are created species of time, which we unconsciously but wrongly transfer to the eternal essence; for we say indeed that he was, he is, he will be, but the truth is that "he is" alone truly expresses him, and that "was" and "will be" are only to be spoken of generation in time, for they are motions, but that which is immovably the same cannot become older or younger by time, nor ever did or has become, or hereafter will be, older, nor is subject at all to any of those states of generation which attach to the movements of sensible things. These are the forms of time when imitating eternity and moving in a circle measured by number. Moreover, when we say that what has become has become, and what is becoming is becoming, and that what will become will become, and that what is not is not, all these are inaccurate modes of expression. But perhaps this is not the place in which we should discuss minutely these matters.

Time, then, was created with the heaven, in order that being produced together they might be dissolved together, if ever there was to be any dissolution of them; and was framed after the pattern of the eternal nature, that it might, as far as possible, resemble it, for that pattern exists throughout all ages, and the created heaven has been, and is, and will be in all time. Such was the mind and thought of God in the creation of time. And in order to accomplish this creation, he made the sun and moon and five other stars, which are called the planets, to distinguish and preserve the numbers of time, and when God made the bodies of these several stars he gave them orbits in the circle of the other. There were seven orbits, as the stars were seven; first, there was the moon in the orbit nearest the earth, and then the sun in the next nearest orbit beyond the earth, and the morning star and the star sacred to Hermes, which revolve in their orbits as swiftly as the sun, but with an opposite principle of motion, which is the reason why the sun and Hermes and Lucifer meet or overtake, and are met or overtaken by each other. To enumerate the places which he assigned to the other stars, and the reasons of them, if they were all to be counted, though a secondary matter, would give more trouble than the primary ones. These things at some future time, when we are at leisure, may have the consideration which they deserve, but not at present.

Now, when all the stars which were needed to make time had attained a motion suitable to them, and their bodies fastened by vital chains, had come into being as living creatures, and learnt their appointed task according to the motions of the other, which is oblique, and passes through and is overruled by the motions of the same, they revolved, some in a larger and some in a lesser orbit, those which have the lesser orbit revolving faster, and those which have the larger moving more slowly. But in the movement of the same, those which revolved fastest appeared to overtake and be overtaken by those which moved slower; for all the orbits of the same moved in a spiral, because they went two ways in opposite directions, and hence that which receded most slowly from the sphere of the same, which was the swiftest, appeared to follow it most nearly. That there might be some visible measure of their relative swiftness and slowness as they proceeded in their eight courses, God lighted a fire, which we now call the sun, in the second of these orbits, that it might give light to the whole of heaven, and that the animals, who were by nature fitted, might participate in number: this was the lesson which they were to learn from the revolutions of the same and the like. Thus, then, and by these means the night and the day were created, being the period of the one most intelligent revolution. And the month was created when the moon had completed her orbit and overtaken the sun, and the year when the sun had completed his own orbit.

39

The periods of the other stars have not been understood by men in general, but only by a few, and they have no name for them, and do not estimate their comparative length by the aid of number, and hence they are hardly aware that their wanderings, which are infinite in number and admirable for their variety, make up time. And yet there is no difficulty in seeing that the perfect number of time completes the perfect year when all the eight revolutions, having their relative degrees of swiftness, are accomplished together and again meet at their original point of departure, measured by the circle of the same moving equally. Thus, and to this end, came into existence such of the stars as moved and returned through the heaven, in order that the created heaven might be as like as possible to the perfect and intelligible animal, and imitate the eternal nature.

Until the creation of time, all things had been made in the likeness of that which was their pattern, but in so far as the universe did not as yet include within itself all animals, there was a difference. This defect the Creator supplied by fashioning them after the nature of the pattern. And as the mind perceives ideas or species of a certain nature and number in the ideal animal, he thought that this created world ought to have them of a like nature and number. There are four such; one of them

40 is the heavenly race of the gods; another, the race of birds moving in the air; the third, the watery species; and the fourth, the pedestrian and land animals. Of the divine, he made the greater part out of fire, that they might be the brightest and fairest to the sight, and he made them after the likeness of the universe in the form of a circle, and gave them to know and follow the best, distributing them over the whole circumference of the heaven, which was to be a true cosmos or glory spangled with them. And he bestowed on each of them two motions; first, the motion in the same, because they ever continue thinking about the same things, and also a forward motion, in that they are controlled by the revolution of the

same and the like; but the other five motions were wanting in them and thus each of them was the best possible. And for this reason also the fixed stars were created, being divine and eternal animals, ever-abiding and revolving after the same manner and on the same spot; and the other stars which revolve and also wander, as has been already described, were created after their likeness. The earth, which is our nurse, compacted (or *circling*) around the pole which is extended through the universe, he made to be the guardian and artificer of night and day, first and eldest of gods that are in the interior of heaven. Vain would be the labor of telling about all the figures of them moving as in a dance, and their meetings with one another, and the return of their orbits on themselves, and their approximations, and to say which of them in their conjunctions meet, and which of them are in opposition, and how they get behind and before one another, and at what times they are severally eclipsed to our sight and again reappear, sending terrors and intimations of things about to happen to those who can calculate them—to attempt to tell of all this without looking at the models of them would be labor in vain. Let what we have said about the nature of the created and visible gods be deemed sufficient and have an end.

To tell of other divinities, and to know their origin, is beyond us, and we must accept the traditions of the men of old time who affirm themselves to be the offspring of the gods, and they must surely have known the truth about their own ancestors. How can we doubt the word of the children of the gods? Although they give no probable or certain proofs, still, as they declare that they are speaking of family traditions, we must believe them in obedience to the law. In this manner, then, according to them, the genealogy of these gods is to be received and narrated:

Oceanus and Tethys were the children of Earth and Heaven, and from these sprang Phorcys and Cronos and Rhea, and many more with them; and from Cronos and Rhea sprang

41

Zeus and Hera, and all those whom we know as their brethren, and others who were their children.

Now, when all of them, both those who visibly appear in their revolutions as well as those other gods who are of a more retiring nature, had come into being, the Creator of the universe spoke as follows: "Gods and sons of gods who are my works, and of whom I am the artificer and father, my creations are indissoluble, if so I will. All that is bound may be dissolved, but only an evil being would wish to dissolve that which is harmonious and happy. And although being created, ye are not altogether immortal and indissoluble, ye shall certainly not be dissolved, nor be liable to the fate of death; having in my will a greater and mightier bond than those which bound you when ye were created. And now, listen to my instructions: Three tribes of mortal beings remain to be created, without them the universe will be incomplete, for it will not have in it every kind of animal which a perfect world ought to have. On the other hand, if they were created and received life from me, they would be on an equality with the gods. In order then that there may be mortals, and that this universe may be truly universal, do ye, according to your natures, betake yourselves to the formation of animals, imitating the power which I showed in creating you. The divine and immortal part of them, which is the guiding principle of those who are willing to follow justice and the gods— of that divine part I will myself give you the seed and beginning. And do you then weave together the mortal and immortal, and make and beget living creatures, and give them food, and make them to grow, and receive them again in death." Thus he spake, and once more and in the same manner poured the remains of the elements into the cup in which he had previously mingled the soul of the universe, no longer, however, pure as before, but diluted to the second and third degree. And when he had framed the universe he distributed souls equal in number to the stars, and assigned each soul to a star; and having placed them as in a chariot, he showed them the nature of the universe, and the decrees of destiny appointed for them, and told them that their first birth would be one and the same for all, and that no one should suffer at his hands; and that they must be sown in the vessels of the times severally adapted to them, and then there would come forth the most religious of animals; and as human nature was of two kinds, the superior race 42 would hereafter be called man. Now, as they were implanted in bodies by necessity, and objects were always approaching or receding from them, in the first place there was a necessity that they should have one natural mode of perceiving external force; in the second place, they must have love, which is a mixture of pleasure and pain; also fear and anger, and the feelings which are akin or opposite to them; if they conquered these they would live righteously, and if they were conquered by them, unrighteously. Also, he said, that he who lived well during his appointed time would return to the habitation of his star, and there have a blessed and suitable existence. But if he failed in attaining this, in the second generation he would pass into a woman, and should he not cease from evil in that condition, he would be changed into some brute who resembled him in his evil ways, and would not cease from his toils and transformations until he followed the original principle of sameness and likeness within him, and overcame, by the help of reason, the later accretions of turbulent and irrational elements composed of fire and air and water and earth, and returned to the form of his first and better nature. When he had given all these laws to his creatures, that he might be guiltless of their future evil, he sowed some of them in the earth, and some in the moon, and some in the other stars which are the measures of time; and when he had sown them he committed to the younger gods the fashioning of their mortal bodies, and desired them to furnish what was still lacking to the human soul, and make all the suitable additions, and rule and pilot the mortal animal in the best and wisest manner that they could, and avert all but self-inflicted evils.

When the Creator had given all these commands he remained in his own nature, and his children heard and were obedient to their father's command, and receiving from him the immortal principle of a mortal creature, in imitation of their own creator they borrowed portions of fire, and earth, and water, and air from the world which were hereafter to be returned; these they took and welded together, not with the indissoluble chains by which they were bound themselves, but with numerous pegs invisible on account of their smallness which they welded together, forming out of them all one body which was subject to influx and efflux, and fastened the courses of the immortal soul in the body. Now these courses, detained as in a vast river, neither overcame nor were overcome; but bore and were borne along violently, so that the whole animal was moved and progressed, irregularly however and irrationally, and in any direction, wandering and coursing according to the six kinds of motion backwards and forwards, and right and left, and up and down, and every way according to the six directions of place. For great as was the advancing and retiring flood which provided nourishment, the affections produced by external contact caused still greater tumult—when the body met and came into collision with external fire, or with the solid earth or the gliding waters, or was caught in the whirlwind hurried along by the air, and the motions produced by any of these impulses were carried through the body to the soul. All such motions have consequently received the general name of "sensations," which they still retain. And these at the moment occasion a very great and mighty movement; they accompany the ever-flowing current, and stir and shake the courses of the soul, and altogether bind fast the course of the same with their opposing flux and hinder it from ruling and proceeding; and the nature of the other they so shake, that the three intervals which formed a progression of doubles (1, 2, 4, 8) and also of triples (1, 3, 9, 27), together with the mean terms and connecting links of the ratios of 3:2, and 4:3, and of 9:8, which cannot be wholly dissolved except by him who tied them together, they twist in all sorts of ways, and bend and disorder the circles as far as they can, so that they are tumbling to pieces, and move irrationally, at one time in the opposite direction, and then again obliquely, and then upside down, as you might imagine a person who is upside down and has his head leaning upon the ground and his feet up against something in the air; and when he is in such a condition, both he and the spectators fancy that the right of the other is his left, and the left right. This and the like of this is what violently moves the courses when they meet with some external thing, either of the class of the same or of the other; and they speak of it as the same with something, or the other of something in a manner which is the very opposite of the truth; and they become false and foolish, and there is no course or revolution in them which has a guiding or directing power; and if again any sensations enter in violently from without and drag after them the whole vessel of the soul, then though they seem to conquer they are really conquered.

44

And by reason of all these affections, the soul when inclosed in a mortal body is at first without sense; but when the stream of growth and nutriment flows in with diminished speed, and the courses of the soul attaining a calm go their own way and become steadier as time advances, then the revolutions of the several circles return to their natural figure, and call the same and the other by their right names, and make the possessor of them a rational being. And if these combine in him with any true nurture or education, he attains the fullness and health of the perfect man, and escapes the worst disease of all; but if he neglects education he walks lame throughout existence in this life, and returns imperfect and good for nothing to the world below. This, however, is an after-stage, and our business now is to treat more accurately of our present subject. There are previous matters relating to the generation of the body and its members, and as to how the soul was created, and from what causes and by what foreknowledge of the gods, which have to be discussed; in this inquiry we hold fast to probability, that is the path in which we must proceed.

First, then, the gods, imitating the spherical shape of the universe, inclosed the two divine courses in a spherical body, that, namely, which we now term the head, being the most divine part of us and the lord of all that is in us: to this the gods who put together the body gave all the rest to be a servant, contriving that it should partake of every sort of motion; in order then that it might not tumble about among the deep and high places of the earth, but might be able to get out of the one and over the other, they provided the body to be a vehicle and means of locomotion; which consequently had length and was furnished with four limbs extended and jointed; these the gods contrived as instruments of locomotion with which it might take hold and find support, and so be able to pass through all places, carrying on high the dwelling-place of the most sacred and divine part of us. This was the origin of legs and arms, which were therefore attached to all men; and the gods, esteeming the front part of man as more honorable and having more authority than the hinder part, they gave men mostly a forward motion. Now it was necessary that man should have his front part distinguished and unlike the rest of his body. Wherefore also about the vessel of the head, in the first place they put in a face in which they inserted organs to minister in all things to the providence of the soul, and they assigned to this anterior part a share of authority. And of the organs they first contrived the eyes to give light, fixing them by a cause on this wise: they contrived that as much of fire as would not have the power of burning, but would only give a gentle light, the light of everyday life, should be formed into a body; and the pure fire which is within us and akin

45

to this they made to flow through the eyes in a single, entire, and smooth substance, at the same time compressing the center of the eye so as to retain all the denser element, and only to allow this to be sifted through pure. When therefore the light of day surrounds the stream of vision, then like falls upon like, and there is a unison, and one body is formed by natural affinity according to the direction of the eyes, wherever the light that falls from within meets that which comes from an external object. And everything being affected by likeness, whatever touches and is touched by this stream of vision, their motions are diffused over the whole body, and reach the soul, producing that perception which we call sight. But when the external and kindred fire passes away in a night, then the stream of vision is cut off; for going forth to the unlike element it is changed and extinguished, being no longer of one nature with the surrounding atmosphere which is now deprived of fire: and the eye no longer sees, and we go to sleep; for when the eyelids are closed, which the gods invented as the preservation of the sight, they keep in the internal fire. And the power of the fire diffuses and equalizes the inward motions; and when they are equalized there is rest, and when the rest is profound, sleep comes with few dreams; but wherever the greater motions $\quad$ 46 remain, whatever may be their nature and situations, they engender corresponding visions within us, and which are remembered by us when we are awake and in the external world. And now there is no longer any difficulty in understanding the creation of images in mirrors and in all smooth and bright surfaces. The fires from within and from without communicate about the smooth surface, and form one image which is variously refracted. All which phenomena necessarily arise by reason of the fire or light about the face combining with the fire or ray of light about the smooth and bright surfaces. And when the parts of the light within and the light without meet and touch in a manner contrary to the usual mode of meeting, then the right appears to be left and the left right; but the right again appears right, and the left left, when the position of one of the two concurring lights is inverted; and this happens when the smooth surface of the mirror, which is convex, repels the right stream of vision to the left side, and the left to the right.[4] Or if the mirror be turned long-ways, then the face appears upside down, and the upper part of the rays are driven downwards, and the lower upwards.

These are the works of the second and coöperative causes which God uses as his ministers when executing the idea of the best as far as possible. They are thought by most men not to be the second, but the prime causes of all things, which they cool and heat, and contract and dilate, and the like. But [this is not true, for] they are incapable of reason or intellect; the only being which can properly have mind is the soul, and this is invisible; whereas fire and water, and earth and air, are all of them visible bodies. Now the lover of intellect and knowledge ought to explore causes of intelligent nature first of all, and, sec-ondly, those which are moved of others and of necessity move others. And this is what we also must do. Both kinds of causes should be considered by us, but a separation should be made of those which are endowed with mind and are the workers of things fair and good, and those which are deprived of intelli-gence and accomplish their several works by chance and with-out order. Of the second or concurrent causes of sight, which give the eyes the power which they now possess, enough has been said. I will therefore proceed to speak of the higher use and purpose for which God has given them to us. The sight in my opinion is the source of the greatest benefit to us, for had the eyes never seen the stars, and the sun, and the heav-en, none of the words which we have spoken about the universe would ever have been uttered. But now the sight of day and night, and the revolution of the months and years, have given us the invention of number, and a conception of time, and the power of inquiring about the nature of the whole; and from this

47

source we have derived philosophy, than which no greater good ever was or will be given by the gods to mortal man. This, I say, is the greatest boon of sight: and of the lesser benefits why should I speak, the loss of which even the common mind would vainly bewail? Thus much let us say: that God invented and gave us sight to this end, that we might behold the courses of intelligence in the heaven, and apply them to the courses of our own intelligence which are akin to them, the unperturbed to the perturbed; and that we, learning them and being partakers of the true computations of nature, might imitate the absolutely unerring courses of God and regulate our own vagaries. The same may be affirmed of speech and hearing; they have been given by the gods for the same ends and purposes. For speech greatly contributes to this purpose, and this is the chief use of musical sound, which is given to the hearing for the sake of harmony. And harmony, which has motions akin to the revolutions of our souls, is not regarded by him who intelligently uses the Muses as given by them with a view to irrational pleasure, which is the prevailing opinion, but with a view to the inharmonical course of the soul, and as an ally for the purpose of reducing this into harmony and agreement with itself; and rhythm was given by them for the same purpose, on account of the irregular and graceless ways which prevail among mankind generally, and to help us against them.

Thus far in what we have been saying, with small exceptions, the works of intelligence have been set forth; and now we must place by the side of them the things done from necessity—for the creation is mixed, and is the result of a union of necessity and mind. Mind, the ruling power, persuaded 48 necessity to bring the greater part of created things to perfection, and thus in the beginning, when the influence of reason got the better of necessity, the universe was created. But if a person will truly tell of the way in which this came to pass, he must include the other influence of the variable cause as well. Wherefore, we must return again and find another suitable beginning, as about the former matters, so also about these. To which end we must consider the nature of fire, and water, and air, and earth, which were prior to the generation of the heavens, and what happened before them; for no one has as yet explained their generation, but we speak of fire and the rest of them, whatever they mean, as though men knew their natures, and we maintain them to be the letters or elements of the whole, when they cannot reasonably be compared by a man of any sense even to the syllables or first compounds. And let me say thus much: I will not speak of the first principle or principles of all things, or by whatever name they are to be called, for this reason, because it is difficult to set forth my opinion according to the mode of discussion which we are at present employing. Do not imagine, anymore than I can bring myself to imagine, that I should be right in undertaking so difficult a task. I will observe the rule of probability with which I began, and I will do my best to speak probably; and, above and before all[5] at the beginning of each and all. Once more, then, I call upon God, at the beginning of my discourse, and beg him to be our savior out of a strange and unwonted inquiry, and to bring us to probability. And now let us begin again.

This new beginning of our discussion of the universe requires a fuller division than the former; for then we made two classes, now a third must be added. For those two classes were sufficient for the former discussion: one which was assumed by us to be a pattern intelligible and always the same; and there was a second, which was only the imitation of the pattern, generated and visible; the third kind we did not distinguish at the time, conceiving that the two would be enough. But now the argument seems to require that we should make clear another kind, which is difficult of explanation and dimly seen. What natural power are we to attribute to this new kind of being? Such a power as this, that it is the receptacle, and in a manner

49

the nurse, of all generation. I have said the truth; but I must give a clearer explanation, and this will be an arduous task for many reasons, and in particular because I must first raise questions concerning fire and the other elements, and say what each of them is; for example, which of them is properly called water as distinct from fire, and by what name any element is called as distinguished from each and all of them; and to give a certain or satisfactory proof of this is not easy. How, then, and in what way, can we arrive at any probable conclusion in this difficulty?

In the first place, that which we are now calling water, when congealed becomes stone and earth, as our sight seems to show us; and this same element, when melted and dispersed, passes into vapor and air. Air, again, when burnt up, becomes fire; and again fire, when condensed and extinguished, passes once more into the form of air; and once more, air, when collected and condensed, produces cloud and vapor; and from these, when still more compressed, comes flowing water, and from water comes earth and stones once more; and thus generation appears to be transmitted from one to the other in a circle. Thus, then, as the elements never appear in the same form, how can anyone have the assurance to maintain strongly that any of them is one thing rather than another? No one can. But much the safest plan is to speak of them as follows: Let us not call that which we see to be continually changing "fire," but rather say, "that some such nature is fire"; and let us not speak of that other thing as water, but rather say that some such nature is water; and let us not speak of objects at all as having stability or erroneously imagine ourselves to indicate any of them by the term "this" or "that," for they are too volatile to be detained in any such expressions as this, or the nature of this, or the nature belonging to this, or any other form of language which implies their permanence. We must not speak of them as individual things, but rather say, of each find all of them, that there is some such uniform principle which circulates in them; for example, of fire we should say that

the general principle is of such a nature always, and so of everything that has generation. The place in which these principles
severally grow up, and appear, and decay, that alone is to
be called by the name "this" or "that"; but that which is of a
certain nature, hot or white, or their opposites, and all that
proceeds from them, are not to be so denominated. Let me
make one more attempt to explain my meaning more clearly.
Suppose a person to make all kinds of figures of gold, and never
to cease transforming out of one form into all the others; some-
body points to one of them and asks, What is that? By far the
safest and truest answer is, That is gold; and not to speak of
the triangle or of any other figures which are formed in the
gold as having real existence, inasmuch as they are in process
of change while he is making the assertion; but if he be willing
to take the general answer, it is enough. And the same may be
said of the universal nature which receives all bodies—that
must be always called the same; for, while receiving all things,
she never departs at all from her own nature, and never in any
way, or at any time, assumes a form like that of the things which
enter into it, being in fact the natural recipient of all impressions, which is moved and fashioned by them, and varies in
appearance from time to time because of them. Now the images
of realities which enter in and go out are modeled after their
patterns in a wonderful and inexplicable manner, which shall
be hereafter investigated by us. But for the present we have only
to conceive of three natures: first, that which is in process of
generation; secondly, that in which the generation takes place;
and thirdly, that of which the thing generated is the natural
resemblance. Moreover, we may liken the receiving principle to
a mother, and the source or spring to a father, and the intermediate nature to a child; and may remark further, that if the model
is to take every variety of form, then the matter in which the
model is fashioned, when duly prepared, must be formless,
and the forms must come from without. For if the matter were

50

like any of the supervening forms, then when any opposite or entirely different nature was impressed the representation would be a bad one, because the matter would shine through. Wherefore, that which is to receive all forms should have no form; as in making perfumes they first contrive that the liquid substance which is to receive the scent shall be as inodorous as possible. Or as those who wish to impress figures on soft substances do not allow any previous impression to remain, but make the surface as even and smooth as possible. 51 In the same way that which is to receive perpetually and through its whole extent the resemblances of eternal beings ought to be destitute of any particular form. Wherefore, the mother and receptacle of all created and visible, and in any way sensible things, is not to be termed earth, or air, or fire, or water, or any of their compounds, or any of the elements out of which they are composed, but is an invisible and formless being which receives all things and attains in an extraordinary way a portion of the intelligible, and is most incomprehensible. In saying this we shall not be far wrong; as far, however, as we can attain to a knowledge of her from the previous considerations, we may truly say that fire is that part of her nature which is inflamed, and water that which is moist, earth and air being also parts, as far as the mother substance receives the impressions of them.

Let us consider this question more precisely. Is there any self-existent fire? And are all those things of which we speak self-existent? Or are only those things which we see, or in some way perceive through the bodily organs, truly existent, and no others besides them? And is all that which we call an intelligible essence nothing at all and only a word? Here is a question which we must not leave unexamined or undetermined, or affirm too confidently that there can be no decision; neither must we interpolate in our present long discourse a digression as long, but if there be a way in which a great principle may be set forth in a few words, that will be just what we want.

Thus I state my view: If mind and true opinion are two distinct classes, then I say that there certainly are these self-existent ideas unperceived by sense, and apprehended only by the mind; but if, as some say, true opinion differs in no respect from mind, then everything that we perceive through the body is to be considered as most real and certain. But we must affirm them to be distinct, for they have a distinct origin and are of a different nature, and the one is implanted in us by instruction, and the other by persuasion, and the one is always accompanied by true reason, and the other is without reason; the one is not to be moved by persuasion, but the other may be moved; and lastly, every man may be said to share in the one but mind is shared only by the gods and by very few men. Wherefore, also, we must acknowledge that there is one kind of being which is always the same, uncreated and indestructible, never receiving anything into itself from without, nor itself going out to any other, but invisible and imperceptible by any sense, and of which the sight is granted to intelligence only. And there is another nature of the same name with it, and like to it, perceived by sense, generated, always in motion, becoming in place and again vanishing out of place, which is apprehended by opinion and sense. And there is a third nature, which is space, and is eternal, and admits not of destruction, and provides a home for all created things, and is perceived without the help of sense, by a kind of spurious reason, and is hardly a matter of belief, which we behold as in a dream, and say, that all existence must of necessity be in some place and occupy a space, and that what is neither in heaven nor in earth has no existence. These things, and others akin to these, relating to the true and waking reality of nature, we, having only this dreamlike sense of them, are unable to arouse ourselves truly to describe or to determine. For an image, not possessing that of which the image is, and existing ever as the changing shadow of some other, must for this reason be in another [i.e., in space], and in

52

some way take hold of essence, or not be at all. But true and exact reason vindicating the nature of true being, maintains that while two things (i.e., the idea and the image) are different they cannot exist one of them in the other so as to become one and also two at the same time.

Thus have I concisely given the result of my thoughts; and my opinion is that being and space and generation, these three, in their three manners existed before the heaven; and that the nurse of generation, moistened by water and inflamed by fire, and receiving the various forms of earth and air, and experiencing all the other accidents that attach to them, took a variety of shapes; and being full of powers which were neither similar nor equally balanced, was never in any part in a state of equipoise, but swaying unevenly to and fro, was shaken by them, and by its motion again shook them, and the elements when moved were divided like the grain shaken and winnowed by fans and other instruments used in the threshing of corn, when the close and heavy particles are borne away and settle in one direction, and the loose and light particles in another. In this manner the four kinds or elements were then shaken by the recipient matter which was itself moved, and like a winnowing machine separated off the elements most unlike from one another, and thrust the similar elements together. Wherefore also these had different places before the universe that was arranged out of them came into being. And at first all things were without reason and measure. But when the world began to get into order, first fire and water and earth and air, having only certain faint traces of themselves, and being altogether such as everything may be expected to be in the absence of God—this, I say, being their nature, God fashioned them by form and number. Let us always, and in all that we say, hold that God made them as far as possible the fairest and best, out of things which were not fair and good. And now I will endeavor to show you the disposition and generation of them by an unaccustomed argument,

53

which however you will be able to follow, for the methods which I must use will be those with which your education has made you familiar.

In the first place, then, as is evident to all, fire and earth and water and air are bodies. And every sort of body possesses solidity, and every solid must necessarily be contained in planes; and the plane rectilinear figure is composed of triangles; and all triangles are originally of two kinds, both of which are made up of one right and two acute angles; one of them has at either end of the base the half of a right angle which is divided by equal sides, while in the other unequal parts of a right angle are divided by unequal sides. These, then, we assume to be the original elements of fire and the other bodies, as we affirm, proceeding by a combination of probability with demonstration; but the principles which are prior to these God only knows, and he of men whom God loves. And next we have to determine what are the four most beautiful bodies which are unlike one another, and yet in some instances capable of resolution into one another; and when we have discovered this, we shall know the true origin of earth and fire and the proportionate and intermediate elements. And then we shall not be willing to allow that there are visible bodies fairer than these, having distinct kinds. Wherefore we must endeavor to join together these four forms of bodies which excel in beauty, and be able to say that we have sufficiently apprehended their nature. Of the two triangles, the isosceles has one form only; the scalene or unequal-sided has an infinite number. Of the infinite forms we must select the most beautiful, if we are to proceed in due order. But if anyone can show a more beautiful form for the composition of these bodies, he shall carry off the palm, not as an enemy but as a friend. Now, the one which we maintain to be the most beautiful of all the many figures of triangles (and we need not speak of the others) is that of which the double forms an equilateral triangle; the

54

reason of this would be long to tell; he who disproves the fact
and proves that this is otherwise is entitled to a friendly victory.
Then let us choose two triangles, out of which fire and other
bodies have been constructed, the one isosceles, the other having
a longer side, the square of which is three times as great as
the square of the lesser side.

Now is the time to explain what was before obscurely said:
there was an error in imagining that all the four elements might
be generated by and into one another; this, I say, was wrong,
for there are generated from the triangles which we have taken
four kinds—three from the one which has the sides unequal;
the fourth alone is framed out of the isosceles triangle. Hence
they cannot all be resolved into one another, or compounded
into larger out of smaller bodies, or the reverse. But three of
them can be thus resolved and compounded, for they all spring
from one, and when the greater bodies are dissolved, many small
bodies will spring up out of them and take their own proper
figures; or, again, when many small bodies are distributed in
triangles, a single number will unite them into one large mass of
another kind. So much for their passage into one another. I
have now to speak of their several kinds, and show out of what
combinations of numbers each of them was formed. The first
kind will be that which is smallest, and its element is that triangle
which has its hypothenuse twice the lesser side. When two such
triangles are joined at the diagonal, and this is repeated three
times, and the triangles rest their diagonals and shorter sides on
the same point as a center, a single equilateral triangle is formed
out of six triangles, and four equilateral triangles, if put
together, make out of every three plane angles one solid
angle [= two right angles], which is nearest to the most
55
obtuse of plane angles; and out of the combinations of these
four angles arises the first solid form which distributes into
equal and similar parts the whole surface. The second species of
solid is formed out of the same triangles, which unite as eight

equilateral triangles and form one solid angle out of four plane angles, and out of six such angles the second body is completed. And the third body is made up of 120 triangular elements, forming twelve solid angles, each of them included in five plane equilateral triangles, having altogether twenty bases, each of which is an equilateral triangle. The one element [that is, the triangle with unequal sides] having generated these figures, generates no more; but the triangle which has equal sides produces the fourth elementary figure, which is compounded of them by fours, joining their right angles in a center, and forming one equilateral quadrangle. Six of these united form eight solid angles, each of which is made by the combination of three plane right angles; the figure of the body thus composed is a cube, having six plane quadrangular equilateral bases. There was yet a fifth combination which God used in the delineation of the universe.

Now, he who, reflecting on all this, inquires whether the worlds are to be regarded as infinite or finite, will be of opinion that the notion of their infinity is characteristic of a very indefinite and ignorant mind. There is, however, more reason in doubting whether they are to be truly regarded as one or five. My opinion is that they are one, and this I deem probable; another, regarding the question from another point of view, may be of another mind. But, leaving this inquiry, let us proceed to distribute the elementary forms, which have now been created in idea, among the four elements.

To earth, then, let us assign the cubical form; for earth is the most immovable of the four and the most easily modeled of all bodies, and that which has the most stable bases must of necessity be of such a nature. Now, of the triangles which we mentioned at first, that which is of equal sides is by nature more stable than that which has unequal sides; and of the compound figures which are formed out of either, the plane equilateral quadrangle has a more stable and necessary basis than the equilateral triangle, both in the whole and in the parts. Wherefore, in assigning this figure to earth, we adhere to probability; and to water we assign that one of the remaining forms which is the most immovable; and the most movable to fire; and to air that which is intermediate between them. Also we assign the smallest body to fire, and the greatest to water, and the intermediate body to air; and, again, the acutest body to fire, and the next in acuteness to air, and the third to water. Of all these elements, that which has the fewest bases must necessarily be the most movable and the acutest and most penetrating in every direction; and must also be the lightest as being composed of the smallest number of similar particles: and the second body has similar properties in a second degree, and the third body in the third degree. Let it be agreed, then, both according to strict reason and according to probability, that the solid form of the pyramid is the original element and seed of fire; and let us assign the second element in the order of generation to air, and the third to water. We must imagine all these to be so small that no single particle of any of the four kinds is seen by us on account of their smallness: but when many of them are collected together the aggregate is seen. And the ratios of their numbers, motions, and other properties, everywhere the God, as far as necessity consented and allowed, has exactly perfected, and harmonized them all in due proportion.

From all that we have just been saying, the most probable result is as follows: earth, meeting with fire and dissolved by its sharpness, is borne hither and thither, either by dissolution in the fire itself or in the air or in the water, until its parts, meeting together and mutually harmonizing, again become earth, for they can never take any other form. But water, when divided by fire or by air, on reuniting, becomes one part fire and two parts air; and a single volume of air divided becomes two of fire. Again, when a small body of fire is contained in a larger body

of air or water or earth, and both are moving, and the fire strug-
gling is overcome and decomposed, then two volumes of fire
form one volume of air; and when air is overcome and cut up
into small pieces, two and a half parts of air are condensed into
one part of water. Let us consider the matter in this way again.

57 When one of the other elements is fastened upon by the fire,
and is cut by the sharpness of its angles and sides, it coalesces
with the fire, and then ceases to be cut by them any longer.
For among bodies which are similar and uniform, none can
change or be changed by another of the same class and in
the same state. But in the process of transition, and during the
conflict of the weaker with the stronger, the dissolution continues.
Again, smaller bodies detained in larger ones, the few encompassed by the many, which are in process of decomposition and
extinction, only cease from their tendency to extinction when they
consent to pass into the conquering nature, and fire becomes
air and air water. But if one kind of bodies goes and does battle
against bodies of another kind, the process of dissolution continues until they are completely ejected and dissolved, and make
their escape to the kindred element, or else, being overcome
and assimilated to the conquering power, they remain and dwell
with their victors, and from being many become one. And owing
to these affections, all things are changing their place, for the
motion of the receiving principle distributes the multitude of
classes into their natural places; but those things which become
unlike themselves and like other things are hurried by the
concussion into the place of the things which they resemble.

Now all unmixed and primary bodies are produced by these
causes. As to the subordinate species which are included in the
greater kinds, they are to be attributed to the various constitu-
tions of the two original triangles. For these differ in magnitude,
and are larger and smaller and have as many sizes as there are
differences of species. Hence when mingled with themselves
and with one another they are infinite in their diversity, which

those who would arrive at the probable reason of nature ought duly to study.

Unless a person comes to an understanding about the nature and conditions of rest and motion, he will meet with many difficulties in the discussion which follows. Something has been said of this matter already, and something more remains to be said, which is, that motion never exists in equipoise. For to conceive that anything can be moved without a mover is hard or indeed impossible, and equally impossible to conceive that there can be a mover without something that will be moved: motion cannot exist where these are wanting, for these to be in equipoise is impossible, and therefore we assign rest to equipoise and motion to the want of equipoise; and inequality is the cause of the nature which is wanting in equipoise. Of this inequality we have already described the origin. But there still remains the question, why things when divided after their kinds do not cease from motion and transition from one into another; this we will now proceed to explain. The revolution of the universe in which are comprehended all natures, being circular and having a tendency to unite with itself, compresses all things and will not allow any place to be left void. Wherefore, also, fire above all things penetrates everywhere, and air next, as being next in rarity of the elements; and the rest in like manner penetrate according to their degrees of rarity. For those things which are composed of the largest particles have the largest void left in their compositions, and those which are composed of the smallest particles have the least. And the tendency towards condensation thrusts the smaller particles into the interstices of the larger. And thus, when the small parts are placed side by side with the larger, and the lesser divide the greater and the greater unite the lesser, all the elements are borne up and down and every way towards their own places; for the change in the size of each changes their position in space. And these causes generate an

58

inequality which is always maintained, and is continually creating a perpetual motion of the elements in all time.

In the next place we have to consider that there are divers kinds of fire. There are, for example, first, flame; and secondly, those emanations of flame which do not burn but only give light to the eyes; thirdly, the remains of fire, which are seen in things red-hot after the flame has been extinguished. There are similar differences in the air; of which the brightest part is called the ether, as the most turbid sort of air is called mist and darkness; and there are various other nameless kinds which are formed by the inequality of the triangles. Water, again, admits in the first place of a division into two kinds; the one liquid and the other fusile. The liquid kind is composed of the small and unequal particles of water; and moves itself and is moved by other bodies because of the inequality of the particles and the shape of the figure; whereas the fusile kind being formed of large and equal elements is more stable than the other, and is solid and compact by reason of its equability. But when fire gets in and dissolves and destroys the equability, it becomes more movable, and when capable of motion is repelled by the neighboring air and spread upon the earth; and this dissolution of the solid masses is called melting, and the spreading out upon the earth is called flowing. When the fire goes out again it does not pass into a vacuum, but into the neighboring air; and the air which is displaced forces together the liquid and still movable mass into the place which was occupied by the fire, and mingles it with itself. Thus compressed the mass resumes its equability, and is again at unity with itself, because the fire which was the author of the inequality has retreated; and this departure of the fire is called cooling, and the coming together which follows upon it is termed congealment. Of all the kinds termed fusile, that which is the densest and is formed out of the finest and most equable parts is that most precious possession which is called gold, and is hardened by filtration through rock; this is

59

unique in kind, and has a bright and yellow color. A matrix of gold, which is so dense as to be very hard, and is blackened, is termed adamant. There is also another kind which has parts nearly like gold, and of which there are several species; this, which is denser than gold, and contains but a small and fine portion of earth, and is therefore harder, and yet because of the great in terstices within is lighter, is a sort of bright and condensed fluid, and when made into a mass is called brass (?). There is an alloy of earth mingled with it, and when the two parts grow old and are disunited, this comes out in the form of what is called rust. The remaining phenomena of the same kind there will be no difficulty in reasoning out by the method of probabilities. A man may sometimes set aside the arguments about eternal things, and for recreation turn to consider the truths of generation which are probable only; thus he attains a pleasure not to be repented of, and makes for himself during his life a wise and moderate pastime. Let us continue to grant ourselves this indulgence, and recount the series of probabilities which follows next in order.

The water which is mingled with fire being of that sort which is fine and liquid, is called liquid, because of its motion and the way in which it rolls upon the earth; and soft, because its bases give way and are less stable than those of earth. This, when separated from fire and air and isolated, becomes more equable, and by their retirement is compressed into itself; and when thus compressed above the earth is called hail, and when on the earth, ice; and that which is congealed in a less degree and is only half solid, when above the earth is called snow, and when upon the earth, and condensed from dew, hoarfrost. Then, again, there are the numerous kinds of water which have been mingled with one another, and are distilled through plants which grow in the earth; and this class is called by the general name of juices or saps. The unequal admixture of these fluids creates a variety of species: most of which are nameless,

60

but four which are of a fiery nature, are clearly distinguished
and have names. First, there is wine, which warms the soul as
well as the body; secondly, there is the oily nature, which is smooth
and divides the light of vision, and for this reason is bright and
shining and of a glistening appearance, including pitch, the juice
of the castor berry, oil, and other things of a like nature; also,
thirdly, there is the diffusive class, which produce sweetness
extending as far as the passages of the mouth; these are included
under the general name of honey: and lastly there is opium (?),
which differs from all other juices, and is a frothy liquid having
a burning quality which dissolves the flesh.

As to the kinds of earth, that which is filtered through water
passes into stone in the following manner: the water which mixes
with the earth and is broken up in the process, passes into
air, and taking this form mounts into its own place. And as
there is no vacuum the neighboring air is thrust out, and this
being heavy and diffused and coagulated around the mass of
earth, violently compresses it and drives it into the vacant
space from whence the new air had come up; and the earth
when compressed by the air into an indissoluble union with
water becomes rock. The fairer sort is that which is made up
of equal and similar parts and is transparent; that which has
the opposite qualities is inferior. But when all the watery part is
suddenly drawn out by fire, a more brittle substance is formed,
to which we give the name of pottery. Sometimes also the
moisture may remain, and the earth which has been fused
by fire becomes, when cool, a stone of a black color. A like
separation of the water may occur in substances composed
of finer particles of earth, and of a briny nature, and then a
half-solid body is formed, soluble in water—either niter which
is used for purifying oil and earth, or else salt, which harmo-
nizes so well in the combinations of the palate, and is, as the
law testifies, a substance dear to the gods. The compounds
of earth and water are not soluble by water, but by fire only, and

for this reason, neither fire nor air melt masses of earth; this is owing to the smallness of their particles, which enables them easily to penetrate the larger interstices of earth without violence; and they leave the earth unmelted and undissolved, but the particles of water being larger force a passage and dissolve and melt the earth. Earth when not thus consolidated by force is dissolved by water only; when consolidated, by nothing but fire; this is the only body which can find an entrance. The cohesion of water again when very strong is dissolved by fire only—when weaker, then either by air or fire—the former entering the interstices, and the latter penetrating even to the triangles. But nothing can dissolve air when strongly condensed, which does not reach the elements or triangles; or if not strongly condensed, then only fire can dissolve it. As to bodies composed of earth and water, while the water occupies the vacant interstices of the earth and holds them compacted together, the circumfluent particles of water finding no entrance leave the entire mass unaffected; but the particles of fire entering into the interstices of the water, do to the air as the water does to the earth, and are the sole causes of the compound body of earth and water liquefying and becoming fluid. Now these bodies are of two kinds; some of them, such as glass and the fusible sort of stones, have less water than they have earth; on the other hand, substances of the nature of wax and incense have more of water entering into their composition.

61

I have thus set forth the various forms and classes of bodies as they are diversified by their combinations and changes into one another, and now I must endeavor to show how the feelings are produced with which they impress us. In the first place, the bodies which I have been describing are necessarily objects of sense. But we have not yet considered the origin of flesh, or what belongs to flesh, or that part of the soul which is mortal. And these things cannot be explained without also explaining the perceptions of sense; nor can the latter be fully explained without these: and yet to explain them together is hardly possible, for which reason we must explain one first, and then proceed to the other. In order, then, that the inquiry may proceed regularly, let us begin by speaking of the affections which equally concern body and soul.

First, let us see why we say that fire is hot, reasoning from the dividing or cutting power which it exercises on our bodies. We all of us feel that fire is sharp; and we may further consider the fineness of the sides, and the sharpness of the angles, and the smallness of the particles, and the swiftness of the motion; all this makes the action of fire violent and sharp, and enables it to cut whatever it meets. And we must not forget that the original figure of fire [i.e., the pyramid], more than any other form, has a dividing power which cuts our bodies into small pieces, and thus naturally produces that affection to which we give the name of heat, which is derived from this ($\theta\varepsilon\rho\mu\grave{o}\varsigma$, cp. $\theta\varepsilon\rho\acute{\iota}\zeta\omega$, $\kappa\varepsilon\rho\mu\alpha\tau\acute{\iota}\zeta\omega$). Now, the opposite of this is sufficiently manifest, yet for the sake of completeness may here be added. For in the case of moist natures which have to do with the body, the larger particles entering in and driving out the lesser, but not being able to take their places, compress the moist principle in us, which, from being unequal and disturbed, is forced by them into a state of rest and equability, and made to coagulate by pressure. Whereas things brought together contrary to nature are naturally at war, and repel one another; and to this war and convulsion the name of shivering and trembling is given; and the whole affection and the cause of the affection are both termed cold. That is called hard to which our flesh yields, and soft which yields to our flesh; and things are also termed hard and soft relatively to one another. That which yields has a small base; but that which rests on quadrangular bases is firmly posed and offers the greatest resistance, and is also that which is the most compact and therefore repellent. The nature of the light and the heavy will be best understood when examined in

62

connection with our notions of above and below; for it is quite wrong to suppose that the universe is parted into two regions, separate from and opposite to each other, the one a lower one to which all things tend which have any bulk, and an upper one to which things only ascend against their will. For as the universe is a globe, all the extremities being equidistant from the center are equally extremities, and the center which is equidistant from them is equally to be regarded as the opposite of them all. Such being the nature of the world, when a person says that anything is above and below, may he not justly be charged with using an improper expression? For the center of the world cannot be rightly called either above or below, but is the center and nothing else; and the circumference is not the center, and has in no one part a greater tendency to the center than in any of the opposite parts. Indeed, when the parts are in every direction similar, how can one rightly give them names which imply opposition? For if there were any solid body in 63 equipoise at the center of the universe, it could not be carried to any of the extremes on account of their perfect similarity; and if a person were to go round it in a circle, he would often, when standing at the antipodes, speak of the same as above and below: for, as I was saying just now, to speak of the whole which is in the form of a globe as having one part above and another below is not like a sensible man. The reason why these terms are used, and the cases in which they are ordinarily applied by us to the division of the heavens, may be elucidated by the following supposition: If a person were to stand in that part of the universe which is the appointed place of fire, and where there is the great mass of fire to which fiery bodies gather—if, I say, he were to ascend thither, and, having the power to do this, were to abstract particles of fire and put them in scales and weigh them, and then, raising the balance, were to draw the fire by force towards the uncongenial element of the air, it would be very evident that the smaller mass would yield more readily than the larger; for when two things are simultaneously raised by one and the same power, the smaller body must necessarily yield to the superior power with less reluctance than the larger; and the larger body is called heavy and said to tend downwards, and the smaller body is called light and said to tend upwards. And we may detect ourselves who are upon the earth doing precisely the same thing. For we often separate earthy natures, and sometimes we draw the earth itself into the uncongenial element of air by force and contrary to nature, both tending to cling to their native element. But that which is smaller yields to the impulse given by us towards the dissimilar elements more easily than the larger; and the former we call light and the place towards which it is impelled we call above, and the contrary state and place we call heavy and below respectively. These must necessarily differ from one another, because the principal masses of the different elements hold opposite positions; for that which is light in the one place is opposed to that which is light in the other, and the heavy to the heavy, and that which is below to that which is below, and that which is above to that which is above; and in their various states of being and becoming they will all be found to be contrary and transverse and in every way diverse in relation to one another. And about all of them this has to be considered: that the tendency of each towards the kindred elements makes the body which is moved heavy, and the place towards which the motion tends below, and of things which are in a contrary position the contrary is true. Such are the causes which we assign to these phenomena. As to the soft and the rough, everyone who sees them will be able to explain the reason of them to another. For roughness is hardness mingled with inequality, and smoothness is produced by the joint effect of quality and density.

64   The most important of the affections which concern the whole body remains to be considered. This is the cause of pleasure and pain in the things which we have mentioned, and

in all other things which are perceived by sense through the parts of the body, and have pleasures and pains consequent upon them. Let us imagine the causes of every affection, whether of sense or not, to be of the following nature, remembering that we have already distinguished between the nature which moves and that which is immovable; for this is the direction in which we must hunt the prey which we mean to take. A body which is easily moved on receiving any slight impression communicates this to the parts affected, and those to other parts in an ever widening circle, until at last reaching the principle of mind they announce the power of the agent. But a body of the opposite kind, being at rest, and having no circular motion, is alone affected, and does not move any of the neighboring parts; and thus the parts not distributing their first impression to other parts, having no effect of motion on the whole animal, produce no effect on the patient. This is true of the bones and hair and other more earthy parts of the human body; whereas what was said above relates mainly to sight and hearing, because they have in them the greatest force of fire and air. Now, we must conceive of pleasure and pain in this way. An impression produced in us contrary to nature and violent, if sudden, is painful; and, again, the sudden return to nature is pleasant, and that which is gentle and gradual is imperceptible and vice versa. The affection which is easily produced is most readily perceived, and not accompanied by pleasure or pain; as, for example, the affections of sight itself, which has been already said to be in the daytime a body in close union with us; for in the use of sight cuttings and burnings and other affections do not produce pain, nor, again, is there pleasure when the sight returns to its natural state; but the strongest and clearest perceptions are produced in as far as the sense is affected and in proportion as the sight itself meets objects; for there is no such thing as violence either in the composition or division of sight. But bodies which are formed of larger particles yield to the agent only with a struggle; and then they impart their motions to the whole and cause pleasure and pain— pain when alienated from their natural conditions, and pleasure when restored to them. Things which experience gradual withdrawings and emptyings of their nature, and great and sudden replenishments, fail to perceive the emptying, and do perceive the replenishment; these occasion no pain, but the greatest pleasure to the mortal part of the human soul, as is manifest in the case of perfumes. But things which are changed all of a sudden, and only gradually and with difficulty return to their own nature, have all the opposite effects, as is evident in the case of burnings and cuttings of the body.

65

Thus have we discussed the general affections of the whole body, and the names of the agents which produce them. And now I will endeavor to speak of the affections of particular parts, and the causes and agents of them, as far as I am able. In the first place let us add what was omitted when we were speaking of juices, concerning the affections peculiar to the tongue. These, like most of the other affections, appear to be caused by certain compositions and divisions, but they have also more of roughness and smoothness than is found in other affections; for whenever earthy particles enter into the small veins which are the testing instruments of the tongue, reaching to the heart, and fall upon the moist, delicate portions of flesh, when by the process of melting they contract and dry up the little veins, they are astringent if they are rougher, but if not so rough then only harsh. Those of them which are of an abstergent nature, and wash the parts about the tongue, if they do this in excess, and take up into themselves and consume away a part of its nature, like potash and soda, are all termed bitter. Those, again, which are of a weaker sort, and which purge only moderately, are called salt, and have no bitterness or roughness, but are regarded rather as agreeable. Bodies which share in and are softened by the heat of the mouth, and which are inflamed, and again in turn inflame that which heats them, and whose lightness is such that they are carried upwards to the sensations of the head, and cut all that comes in their way, by reason of these qualities in them, are all termed pungent. But when these same particles, refined away by putrefaction, enter into the narrow veins, 66 and there meet the earthy and airy particles, and set them whirling, and while they are in a whirl cause them to interpenetrate with one another and form new hollows exterior to the particles which enter, as happens with the hollow drop surrounding the air, which is sometimes mixed with earth and sometimes pure, in the latter case forming hollow watery vessels of air of a circular shape, pure and transparent, which are called bubbles, while those composed of the earthy liquid which is in a state of general agitation and rising, are called boiling or fermentation, of all these affections the cause is termed acid. And there is the opposite affection arising from an opposite cause, when the composition of the particles which enter dissolved in liquid is congenial to the tongue, and smooths and oils over the roughness, and relaxes the parts which are unnaturally contracted, and contracts the parts which are relaxed, and disposes them all according to their nature; that sort of remedy of violent affections is pleasant and agreeable to every man, and has the name sweet. Enough of this.

As to the faculty of smell, that does not admit of kinds; for all smells are but half-formed substances, and no element is so proportioned as to have any smell. The veins about the nose are too narrow to admit the various kinds of earth and water, and too wide to admit those of fire and air; and for this reason no one ever smells any of them, but smells always proceed from bodies that are damp, or putrefying, or liquefying, or smoking, and are perceptible only in the intermediate state, when water is changing into air and air into water, and all of them are either smoke or mist. That which is passing out of air into water is mist, and that which is passing from water into air is smoke; and hence all smells are thinner than water and thicker than air. The proof of this is, that when there is any obstruction to the respiration, and a man draws in his breath by force, then no smell filters through, but the air only without the smell penetrates; and this is the reason why there are only two varieties of them, because they are not composed of many simple elements, and they have no name, but are distinguished as painful and pleasant, the one irritating and disturbing the whole cavity which is situated between the head and the navel, the other having a soothing influence, and restoring this same region to an agreeable and natural condition.

And now we have to speak of hearing, which is a third kind of sense, and of the causes in which this affection originates. We may assume speech to be a blow which passes through the ears, and is transmitted by means of the air, the brain, and the blood, to the soul, and that hearing is the motion of this blow, which begins in the head and ends in the region of the liver; and the sound which moves swiftly is acute, and the sound which moves slowly is grave, and that which is uniform is equable and smooth, and the reverse is harsh. A great body of sound is loud, and the opposite is low. Respecting the harmony of sounds I must hereafter speak.

There is a fourth class of sensible things, comprehending many varieties, which have now to be distinguished. They are called by the general name of colors, and are a flame which emanates from all bodies and has particles corresponding to the sense of sight. I have spoken already, in what has preceded, of the causes of the generation of sight, and this will be a natural and suitable place in which to give some account of color.

Of the particles coming from other bodies which fall upon the sight, some are less and some are greater, and some are equal to the parts of the sight itself. Those which are equal are imperceptible, or transparent, as they are called by us, whereas the larger contract, the smaller dilate the sight, having a power akin to that of hot and cold bodies on the flesh, or of astringent bodies on the tongue, or of those heating bodies which are termed pungent by us. White and black, although they are found in another class of objects, and for this reason are imagined to be different, are affections of the same kind. Wherefore, we ought to term that white which dilates the visual ray, and the opposite of this black. There is also a swifter motion and impact of another sort of fire which dilates the ray of sight and reaches the eyes, forcing a way through their passages and melting them, and eliciting from them a union of fire and water 68 which we call tears, being itself an opposite fire which comes to them from without—the one flashes forth like lightning, and the other finds a way in and is extinguished in the teardrop, and all sorts of colors are generated in the mixture. This affection we term dazzling, and that which produces it is called bright and flashing. There is another sort of fire which is intermediate, and which reaches and mingles with the moisture of the eye without flashing; and in this, the fire mingling with the ray of the teardrop produces a color like blood, to which we give the name of red. A bright hue mingled with red and white gives the color called auburn (ξανθόν). The law of proportion, however, in which the several colors are formed, even if a man knew he would be foolish if he attempted to tell, as he could not give any necessary reason, nor even any tolerable or probable account of them. Again, red, when mingled with black and white, gives a purple hue, which becomes umber (ὄρφνινον) when the colors are burnt as well as mingled and the black is more thoroughly mixed with them. Flame color (πυρρὸν) is produced by a union of auburn and dun (φαιὸν), and dun by an admixture of black and white; pale yellow (ὠχρὸν) by an admixture of white and auburn. White and light meeting, and falling upon a full black, become dark blue (κυανοῦ), and when dark blue mingles with white, a light blue (γλαυκὸν) color is formed, as leek green (πράσιον) is formed also out of the union of flame color and black. There will be no difficulty in seeing how other colors are to be mingled and assimilated in accordance with probability. He, however, who should attempt to test the truth of them in fact, would forget the difference of the human and divine nature. For God only has the knowledge and also the power which are able to combine many things into one and again dissolve the one into many. But no man either is or ever will be able to accomplish either of these operations.

These are the elements, thus of necessity then subsisting, which the Creator of the fairest and best received in the world of generation, when he made the self-sufficing and most perfect God, using the secondary causes as his ministers in the creation of these things, but himself fashioning the good in all his creations. Wherefore we may distinguish two sorts of causes, the one divine and the other necessary, and may seek for the divine in all things, as far as our nature admits, for the sake of the blessed life; but the necessary kind only for the sake of the divine, considering that without them and when isolated from them, these higher things for which we look cannot be apprehended or received or in anyway attained by us.

69

Seeing, then, that we have now before us the various classes of causes which are the material out of which the remainder of our discourse is to be framed, just as wood is the material of the carpenter, let us recur rapidly to the point at which we began, and then let us endeavor to add on a suitable beginning and ending to our tale.

As I said then at first, when all things were in disorder, God created in each thing, both in reference to itself and to other things, certain harmonies in such degree and manner as they are capable of having proportion and harmony. For in those days nothing had any order except by accident; nor did any of the things which now have names deserve to be named at all—as, for example, fire, water, and the rest of the elements. All these the Creator first arranged, and out of them he constructed the universe, which was a single animal comprehending all other animals, mortal and immortal, in itself. Now of the divine, he himself was the Creator, but committed to his offspring the creation of the mortal. And they, imitating him, received from him the immortal principle of the soul; and around this they fashioned a mortal body, and made the whole body to be a vehicle of the soul, and constructed within a soul of another nature which was mortal, subject to terrible and irresistible affections: first of all, pleasure, the greatest incitement of evil; then pain, which deters from good; also rashness and fear, foolish counselors, anger implacable, and hope easily deceived by sense without reason and by all-daring love: these they mingled together according to necessary laws, and framed man. Wherefore, fearing to pollute the divine anymore than is necessary, they separated the mortal nature, and gave that a habitation in another part of the body, placing the neck between them to be the isthmus and boundary line, which they constructed between the head and the breast, that they might be kept distinct. And in the breast, and in what is 70 termed the thorax or breastplate of man, they encased the mortal soul, and as one part of this was superior and the other inferior they divided the cavity of the thorax into two parts, as the women's and men's apartments are divided in houses; and placed the midriff to be a wall of partition between them. That part of the inferior soul which is endowed with courage and spirit and loves contention they settled nearer the head, in the interval between the midriff and the neck, in order that it might be under the control of reason, and might join with it in forcing and restraining the desires when they are no longer willing of their own accord to obey the command of reason issuing from the citadel.

The heart, which is at once the source[6] of the veins and the fountain of the blood which is in rapid circulation through all the limbs, they placed in the guardhouse, that when the spirit

was roused at the instigation of reason making proclamation
of any wrong assailing them from without or being perpetrated
by the desires from within, quickly the whole power of feeling
in the body, perceiving these commands and threats, might obey
and follow through every turn and alley, and thus allow the principle of the best to have the command in all of them. But as the
gods foreknew that the palpitation of the heart, in the expecta-
tion of danger or in the excitement of anger, was caused by fire,
and that this led to the swelling of passion, they formed and
implanted the lung as a sort of aid to it. Now this was, in the first
place, soft and bloodless, and also had within hollows like
the pores of a sponge, in order that, receiving the breath and the
drink and cooling them, it might give the power of respiration
and alleviate the heat. Wherefore also they cut the passages of
the trachea which lead to the lung, and placed the lung
about the heart as a soft spring, that, when anger was rife
in it, the heart, beating against the yielding body, might be
refreshed and alleviated, and might thus become more ready
to accompany passion in the service of reason.

The part of the soul which desires meats and drinks and
such things as the bodily frame needs, they placed between
the midriff and the navel, contriving in all this region a sort
of manger for the food of the body; and there they bound the
desires down as a wild animal which was chained up with man
and must be nourished if man was to exist. In order that this
lower creature might be always feeding at the manger, and have
his dwelling as far as possible from the council chamber, making
as little noise and disturbance as possible, and permitting the
best part to advise quietly for the good of the whole, they
appointed for him this place. And knowing that this principle in man would not listen to reason, and even if attaining
to some degree of perception would never naturally care for any
arguments, and was liable to be led away by phantoms and visions
of the night and also by day, God, considering this, framed the

71

liver, to connect with the lower nature and to dwell there, contriving that it should be compact and smooth, and bright and sweet, and also bitter, in order that the power of thought, which originates in the mind, might be reflected as in a mirror which receives and gives back images to the sight. And this power, being akin to the bitter part of the liver, by the help of that inspires terror, and comes threatening and invading, and suddenly mingling with the entire liver produces colors like bile, and contracts every part, and makes it wrinkled and rough; or, on the other hand, twisting out of their right place and contracting the lobe and receptacles and gates, or again, closing and shutting them up—in these and other ways creates pain and disgust. And the converse happens when some gentle inspiration of the understanding pictures images of an opposite character, and allays the bile and bitterness by not stirring them, and refuses to touch the nature opposed to itself, but by making use of the natural sweetness of the liver, straightens all things and makes them to be right and smooth and free, and makes the portion of the soul which resides about the liver happy and joyful, having in the night a time of peace and moderation, and the power of divination in sleep when it no longer participates in sense and reason. For the authors of our being, remembering the command of their father when he bade them make the human race as good as they could, thus ordered our inferior parts in order that they too might obtain a measure of truth, and in the liver placed their oracle, which is a sufficient proof that God has given the art of divination to the foolishness of man. For no man, when in his senses, attains prophetic truth and inspiration; but when he receives the inspired word either his intelligence is enthralled by sleep, or he is demented by some distemper or possession. And he who would understand what he remembers to have been said, whether in dream or when he was awake, by the prophetic and enthusiastic nature,

72

or what he has seen, must recover his senses; and then he will be able to explain rationally what all such words and apparitions mean, and what indications they afford to this man or that, of past, present, or future good and evil. But, while he continues demented, he cannot judge of the visions which he sees or the words which he utters; the ancient saying is very true that "only a man in his senses can act or judge about himself and his own affairs." And for this reason it is customary to appoint diviners or interpreters as discerners of the oracles of the gods. Some persons call them prophets; they do not know that they are only repeaters of dark sayings and visions, and are not to be called prophets at all, but only interpreters of prophecy.

Such is the nature and position of the liver, which is intended to give prophetic intimations. During the life of each individual these intimations are plainer, but after his death the liver becomes blind, and delivers oracles too obscure to be intelligible. The spleen is situated in the neighborhood on the left-hand side, and is constructed with a view of keeping the liver bright and pure like a sponge, always ready prepared and at hand to clean the mirror. And hence, when any impurities arise by reason of disorders of the body affecting the liver, the loose nature of the spleen, which is composed of a hollow and bloodless tissue, receives them all and purges them away, and when filled with the unclean matter, becomes enlarged and diseased, but, again, when the body is purged, settles down into the same place as before, and is humble.

Concerning the soul, as to which part is mortal and which divine, and where they exist, and what are their conditions, and why they are separated, the truth can only be established, as has been said, by the word of God; still, we may venture to assume that what has been said by us is probable, and will be rendered more probable by investigation. Let us affirm this.

The creation of the body comes next, and this we may investigate in a similar manner. And it appears to be very meet that the body should be framed on the following principles:

The authors of our race were aware that we should be intemperate in eating and drinking, and take a good deal more than was necessary or proper, by reason of gluttony. In order then that disease might not quickly destroy us, and lest our mortal race should perish and fail of fulfilling its end—intending to provide against this, the gods made a receptacle for the superfluous meat and drink, which is called the lower belly, and formed the convolution of the bowels, so that the food might be prevented from passing quickly through and compel the body to require more food, thus producing insatiable gluttony, and making the whole race an enemy to philosophy and music, and rebellious against the divinest element within us.

73

The bones and flesh, and other similar parts of us, were made as follows: The first principle of all of them was the generation of the marrow. For the bonds of life which unite the soul with the body are made fast there, and they are the root and foundation of the human race. The marrow itself is created out of other elements: God took such of the triangles as were of the first formation, straight and smooth, and specially adapted by their perfection to produce fire and water, and air and earth—these, I say, he separated from their kinds, and mingling them in due proportions with one another, made the marrow out of them to be a universal seed of the whole race of mankind; and after that he planted and inclosed in this the various kinds of souls, and in the original distribution gave the marrow as many and various forms as there were hereafter to be kinds of souls. That which, like a field, was to receive the divine seed, he made round every way, and called that portion of the marrow, brain, intending that, when an animal is perfected, the vessel containing this substance should be the head; but as touching the remaining and mortal part of the soul—that which was intended to contain this—he divided into round and long figures, and he called them all by the name "marrow"; and from these, as from anchors, casting the bonds of the whole soul, he proceeded to fashion around them the entire framework of our body, constructing for the marrow, first of all, a complete covering of bones.

The bones were composed by him in the following manner: Having sifted pure and smooth earth he kneaded it and wetted it with marrow, and after that he put it into the fire and then into the water, and once more into the fire and again into the water—in this way by frequent transfers from one to the other he made it insoluble by either. With this bone he fashioned, as in a lathe, a globe made of bone, which he placed around the brain, and in this globe he left a narrow opening; and around the marrow of the neck and back he formed the vertebrae like hinges, beginning at the head and extending through the whole of the trunk. Thus he preserved the entire seed, which he inclosed in a case like stone, inserting joints, and using in them the intermediate nature of the other, in order to obtain motion and flexion. Then again, considering that the bone would be too brittle and inflexible, and when inflamed and again cooled would soon mortify and destroy the seed within—having this in view, he contrived the sinews and the flesh, that so binding all the members together by the sinews, which admitted of being stretched and relaxed about the vertebrae, he might thus make the body capable of flexion and extension, while the flesh would serve as a protection against the summer heat and against the winter cold, and also against falls, like articles made of felt, softly and easily yielding to external bodies, and containing in itself a warm moisture which in summer exudes in the form of dew, and imparts to the body a natural coolness; and again in winter by the help of its own fire forms a very tolerable defense against external and surrounding cold. The great molder and creator considering this, mingled earth with fire and water and put them together, making a ferment of acid and salt which he mingled with them and formed a soft and pulpy flesh;

74

and as for the sinews, he made them of an unfermented mixture of bone and flesh, attempered so as to be in a mean, and give them a yellow color, and hence the sinews have a firmer and more glutinous nature than flesh, but a softer and moister nature than the bones. With these God covered the bones and marrow, which he bound together with sinews, and then enshrouded them all in an upper covering of flesh. The more living and sensitive of the bones he inclosed in the smallest film of flesh, and those which had the least life he inclosed in the most solid flesh. So again on the joints of the bones, where reason indicated that no more was required, he placed only a small quantity of flesh, that it might not interfere with the flexion of our bodies and make them uneasy because difficult to move; and also that they might not by being crowded and pressed and matted in one another, lose the power of sensation by reason of their hardness, and make the parts which have to do with the mind dull of remembering and hearing. Wherefore also the thighs and the legs and the loins, and the bones 75 of the arms and the forearms and other parts which have no joints, and the inner bones, which on account of the rarity of the soul in the marrow are destitute of reason—all these are filled up with flesh; but such as have feeling are in general less fleshy, except where the Creator has made some part solely of flesh; as, for example, the tongue, in order to give sensation. But generally this is not the case. For the combination of solid bone and much flesh with acute perceptions, is contrary to the laws of the composite nature. More than any other part the framework of the head would have had them, if they could have coexisted, and the human race, having a strong and fleshy and sinewy head, would have had a life twice and many times as long, and also more healthy and free from pain. But our creators considering whether they should make a long-lived race which was worse, or a short-lived race which was better, came to the conclusion that the preference should be given by

everyone to a shorter span of life which was better, rather than
to a longer one which was worse; and therefore they covered the
head which has no flexure with thin bone, but not with flesh and
sinews; and thus the head was added, having more wisdom
and sensation than the rest of the body, but also being in every
man far weaker. And for a like reason God placed the sinews at
the extremity of the head, in a circle round the neck, and glued
them together and fastened the cheeks to them at the extremity
underneath the face, and other sinews he dispersed throughout
the body, fastening limb to limb. The framers of our being
framed the mouth, as now, having teeth and tongue and lips,
with a view to the necessary and the good, contriving the way in
for necessary purposes, the way out for the best purposes; for
that is necessary which enters in and gives food to the body;
but the river of speech which goes out of a man and ministers to
the intelligence is the fairest and noblest of all streams. Still the
head could neither be left a bare frame of bones, on account
of the extremes of heat and cold in the different seasons, nor be
allowed to be wholly covered, and so become dull and senseless
by an overgrowth of flesh. The fleshy nature was not therefore wholly dried up, but a large sort of peel was parted off
and remained over, which is now called the skin. This met and
grew by the help of the cerebral humor, and became the circular envelopment of the head. And the moisture springing up
from beneath the sutures watered and closed them at the top,
fastening them into a knot; the diversity of the sutures was caused
by the power of the courses of the soul and of the food, and
the more these struggled against one another the greater the
diversity became, and grew less if the struggle diminished. This
skin the divine power pierced all round with fire, and out of
the punctures which were thus made the moisture issued forth,
part liquid and hot which came away pure, and a mixed part
which was composed of the same material as the skin, but was
driven upwards and outwards, and extended to a great length,

76

having a fineness equal to the punctures, and being too slow to find an exit, and thrust back by the external air, taking a condensed form, settled underneath the skin. And owing to these affections the hair sprang up in the skin, being of a skinny and stringy nature, but harder and closer through the pressure of the cold, by which each hair separated from the skin is compressed and cooled. In this manner the Creator formed our head all hairy, making use of the causes which I have mentioned, and reflecting also that instead of flesh the part about the brain needed the hair to be a light covering or guard, which would give shade in summer and shelter in winter, and at the same time would not impede our quickness of perception. From the combination of sinew, skin, and bone, in the structure of the finger, there arises a triple compound which, when dried up, fakes the form of one hard skin partaking of all three natures, and was fabricated by these second causes, but designed by the principal mind or cause with an eye to the future. For those who formed us well knew that women and other animals would someday be framed out of men, and they further knew that many animals would require the use of nails for many purposes; wherefore also they stamped in men at their first creation the forms of nails. From this cause and for these reasons they fashioned skin, hair, and nails at the extremities of the limbs.

And now that all the parts and members of the mortal animal had come together, and their life of necessity 77 consisted of fire and spirit, and was liable therefore to melt away and perish from exhaustion, the gods contrived the following remedy for this: they mingled a nature akin to that of man with other forms and perceptions, and thus created another kind of living being. These are the trees and plants and seeds, which by cultivation are now adapted to our use; anciently there were only the wild kinds, which are older than the cultivated. For everything that partakes of life may be truly called a living being, and this of which we are now speaking partakes of the third nature of the soul, which is said to be seated between the midriff and the navel, and has no part in opinion or reason or mind, but only perception of pleasure and pain and the desires which accompany them. For this nature is always in a passive state, and revolving in and about itself, repelling the motion from without and using its own, and not gifted originally with the power of seeing or reflecting on its own concerns. Wherefore it lives and is a living being, but is fixed and rooted in the same spot, having no power of self-motion.

Now, after the superior powers had created all these natures to be food for us who are of the inferior nature, they cut various channels through our bodies, as in a garden, watering them as with a perennial stream. In the first place, they cut two secret channels or veins down the back where the skin and the flesh join, corresponding severally to the right and left side of the body. These they placed along the backbone, so as to receive between them the marrow of generation, the growth of which might be thus promoted, and that the descending flood supplied thence to other parts might equalize the irrigation. In the next place, they divided the veins about the head, and, interlacing them, they sent them in opposite directions; those coming from the right side they sent to the left of the body, and those from the left they turned towards the right, that they as well as the skin might bind the head to the body, inasmuch as the head was not inclosed at the top by the sinews, and also that the sensations from both sides might be distributed over the whole body. And next, they ordered the course of liquids in a manner which I will describe, and which we shall more readily understand if we begin by admitting that all things which are composed of lesser parts retain the greater, but the greater cannot retain the lesser. Now, of all natures fire has the smallest parts, and therefore penetrates through earth and water and air and their compounds, nor can anything hold it; and this is true also of the belly, which is able to

78

retain meats and drinks that are passed into it, but is not able to retain air and fire, which consist of smaller particles than those of which it is composed.

These channels, therefore, God employed for the sake of distributing moisture from the belly into the veins, weaving together a network of fire and air like basket nets, at the entrance of which he made two openings, the one of which he further formed with two branches, and from the openings he extended a sort of cord reaching all round to the extremity of the network. All the inner parts of the network he made of fire, but the openings and the cavity he made of air. The network he took and spread over the newly-farmed animal in the following manner: he let one of the openings pass into the mouth; this opening was twofold, and he let one part of it descend by the air-pipes into the lungs, the other by the side of the air-pipes into the belly. The other opening he divided into two parts, both of which he made to communicate with the channels of the nose, so that when there was no way through the mouth the streams of the mouth were replenished from the nostril. But the other cavity of the network he placed around so much of the body as was hollow, and the entire receptacle which was composed of air he made to flow into the passages of the network, which then flowed back; the tissue of the lung found a way in and out of the pores of the body, and the rays of fire which were interlaced followed the passage of the air either way; this continuing as long as the mortal being holds together. These, as we affirm, are the phenomena which the imposer of names called respiration and expiration. And all this process of cause and effect took place in order that the body might be watered and cooled, and thus have nourishment and life; for when the respiration is going in and out, and the fire, which follows at the same time, is moving to and fro, and, entering through the belly, reaches the meat and drink, it liquefies them, and, dividing them into small portions and guiding them through the

79

passages where it goes, draws them as from a fountain into the channels or veins, and makes the stream of the veins flow through the body as through a conduit.

Let us further consider the phenomena of respiration, and inquire what are the real causes of it. They are as follows. Seeing that there is no such thing as a vacuum into which any of those things which are moved can enter, and the breath is carried from us into the external air, the next point is, as will be clear to everyone, that it does not go into a vacant space, but pushes its neighbor out of its place, and that which is thrust out again thrusts out its neighbor; and in this way of necessity everything at last comes round to that place from whence the breath came forth, and enters in there, and follows with the breath, and fills up the place; and this goes on like the circular motion of a wheel, because there can be no such thing as a vacuum. Wherefore also the breast and the lungs, which emit the breath, are again filled up by the air which surrounds the body and which enters in through the pores of the flesh and comes round in a circle; and, again, the air which is sent away and passes out through the body forces the breath within to find a way round through the passage of the mouth and the nostrils. Now, the origin of this may be supposed to be as follows: Every animal has his inward parts about the blood and the veins as warm as possible; he has within him a fountain of fire, which we compare to the texture of a net of fire extended through the center of the body, while the outer parts are composed of air. Now, we must admit that heat naturally proceeds outward to its own place and to its kindred element; and as there are two exits for the heat, the one through the body outwards, and the other through the mouth and nostrils, when it moves towards the one, it drives round the other, and that which is driven round falls into the fire and is warmed, and that which goes forth is cooled. But when the condition of the heat changes, and the particles at the other exit grow warmer, the hotter air inclining in that

direction and carried towards its native element fire, pushes
round the other; and thus, by action and reaction, there being
this circular agitation and alternation produced by the two, by this
double cause, I say, inspiration and expiration are produced.
The phenomena of medical cupping-glasses and
of the swallowing of drink and of the hurling of bodies,
whether discharged in the air or moving along the ground,
are to be explained on a similar principle; as also the nature
of sounds, whether swift or slow, sharp or flat, which are
sometimes discordant on account of the inequality of the
motion which they excite in us, and then again harmonical on
account of their equality; for the slower sounds reach the
motions of the antecedent swifter sounds when these begin to
pause and come to an equality, and after a while overtake and
propel them. When they overtake them they do not introduce
another or discordant motion, but they make the slower
motion by degrees correspond with the swifter; and when the
motion leaves off, they assimilate them and cause a single
mixed expression to be produced from sharp and flat, whence
arises a pleasure which even the unwise feel, and which
to the wise becomes a higher sort of delight, as being an
imitation of divine harmony in mortal motions. Moreover,
as to the motions of water, the thunderbolt, and the marvels that
are observed about the attraction of amber and the Heraclean
stones, in none of these cases is there any attraction; but, as
there is no vacuum, these substances thrust one another round
and round, all severally passing and succeeding to their own
places by composition and dissolution. Such will appear to
the reasonable investigator to be the causes whose united
influence produces these wonders.

I have spoken of the nature and causes of respiration,
in which our discourse originated. As I before said, the fire
divides the food and rises within in company with the breath;
in the process of respiration filling the veins out of the belly

by drawing from thence the divided portions of the food, by which means the streams of food are diffused through the whole body in all animals. The fruits or grass, which are of a kindred nature, and which God planted to be our daily food, when newly cut, acquire all sorts of colors by reason of their admixture; but the red color for the most part predominates, being a nature made by the cutting power of fire leaving a stain in moisture; and hence the liquid which circulates in the body has such a color as we have described, which we call blood, being the nurturing principle of the flesh and of the whole body, whence all parts are watered and the empty places filled.

81

Now the process of repletion and depletion is effected after the manner of the universal motion of all things, which is due to the tendency of kindred natures towards one another. For the external elements which surround us are always causing us to consume away, and distributing and sending away like to like; the particles of blood, too, which are divided and contained within the frame of the animal, which is a sort of world to them, are compelled to imitate the motion of the universe. Each, therefore, of the divided parts within us, being carried to its kindred nature, replenishes the void. When more is taken away than flows in, then we decay, and when less, we grow and increase.

The young of every animal has the triangles new, and may be compared to the keel of a vessel which is just off the stocks; they are locked closely together and yet the entire frame is soft and delicate, as if freshly formed of marrow and nurtured on milk. Those triangles, therefore, which come in from without and are contained in the bodily frame, from which are formed meats and drinks, being older and weaker than its own triangles, the frame of the body gets the better of them and cuts them up with the new triangles, and the animal grows and is nourished by the assimilation of particles. But when the root of the triangles is relaxed by having undergone many conflicts with many things in the course of time, they are no longer able to cut or assimilate the food which enters into them, but are easily subverted by the new bodies which come in from without. In this way the whole animal is overcome and decays, and this state of things is called old age. But at last, when the bonds of the triangles which inclose the marrow no longer hold, and get unfixed by the toil of which I spoke, they unfix also the bonds of the soul, and she being released, in the order of nature joyfully flies away. For that which is not in the order of nature is painful, but that which takes place according to nature is pleasant. And thus, too, death, if caused by disease or produced by wounds, is painful and difficult; but that sort of death which comes of old age and fulfills the debt of nature is the least painful of deaths, and is accompanied with pleasure rather than with pain.

Now everyone can see whence diseases arise. There are four natures out of which the body is compacted—earth and fire and water and air, and the unnatural excess and defects of these, or the change of any one of them from their own 82 natural place into another, or, again, the assumption on the part of these diverse natures of fire and the like of that which is not suitable to them, or anything of that sort, produces diseases and disorders; for each being produced or changed in a manner contrary to nature, the elements which were previously cool grow warm, and those which were dry become moist, and the light becomes heavy, and the heavy light; all sorts of changes occur. For we affirm that only the same, in the same and like manner and proportion added or subtracted to or from the same, will allow the body to remain in the same state, whole and sound, and that, whatever is taken away or added in violation of these rules causes all manner of changes and infinite diseases and disorders. But as there are secondary compositions which are according to nature, he who will understand diseases may also have another or second notion of them. For whereas the marrow and the bone and the flesh and the sinews are composed of these elements, as the blood is likewise composed of them

but in a different degree, the greater number of them are caused
in the way which I have already mentioned; but the worst
of all owe their severity to the following causes: When the
generation of them proceeds in an order contrary to nature,
then the elements are destroyed. For the natural order is that the
flesh and sinews are made of blood, the sinews out of the fibers
to which they are akin, and the flesh out of the congealed
substance which is formed by separation from the fibers. And
the glutinous and rich matter which comes away from the sinews
and the flesh not only binds the flesh to the bones, but nourishes
and imparts growth to the bone itself which surrounds the
marrow, and by reason of the solidity of the bones, that which
is filtered through is the purest and the smoothest and the oili-
est sort of the triangles which drops like dew from the bones and
waters the marrow. And when these are the conditions, health
usually ensues; when the conditions are of an opposite nature,
disease. For when the flesh becomes melted and sends back the
wasting substance into the veins, then there is a great deal of
blood of different kinds as well as of air in the veins, having
various degrees of color and bitterness: and also from its acid
and salt qualities it generates all sorts of bile and lymph and
phlegm. For all things go the wrong way and are corrupted,
and first of all destroy the blood, and then ceasing to give
nourishment to the body are carried along the veins in all sorts
of ways, no longer preserving the order of their natural courses,
but at war with themselves, because they have no enjoyment of
themselves, and are hostile to the abiding constitution of the
body, which they destroy and waste. The oldest part of the flesh
which wastes away, refusing to assimilate, grows black from
long burning, and from being corroded in every direction
becomes bitter, and is injurious to every part of the body which
is not yet corrupted. And then instead of bitterness the black
part assumes an acidity from the bitter element refining away:
or, again, the bitter substance being tinged with blood has a

83

redder color; or, when mixed with black, has the[7] hue of grass; and once again, an auburn color is mingled with the bitter matter when the new flesh is melted by the fire which surrounds the internal flame; to all which some physician, or some philosopher, who had the power of seeing many dissimilar things and recognizing in them one nature common to them all and deserving of a name, has assigned the common name of bile. But the kinds of bile have also their peculiar names corresponding to their several colors. As for lymph, that sort which is the whey of blood is gentle, but that which is produced by dark and bitter bile is of a fierce nature when mingled by the power of heat with any salt substance, and is then called acid phlegm. Again, the decomposed of new and tender flesh which is accompanied by air when inflated and encased in liquid producing bubbles which separately are invisible owing to their small size, but when collected together are of a bulk which is visible, and have a white color arising out of the generation of foam—all this dissolution of tender flesh when intermingled with air is termed by us white phlegm. And the whey or sediment of phlegm when just formed is sweat and tears, and includes the various secretions which arise daily out of the purgation of the body. Now all these become the instruments of disease when the blood is not replenished according to nature by meats and drinks but gains bulk from contraries in violation of the laws of nature. When the several parts of the flesh are separated by disease, $\quad$ 84 if the foundation remains, the trouble is only half as great, and recovery is still possible; but when that which binds the flesh to the bones is diseased, and the blood, which is made out of the fibers and sinews, separates from them, and no longer gives nourishment to the bone, or is a bond of union to flesh and bone, and from being oily and smooth and glutinous becomes rough and salt and dry, owing to bad regimen, then the substance which is detached crumbles away under the flesh and the sinews, and separates from the bone, and the fleshy parts fall away from their foundation and leave the sinews bare and full of brine, and the flesh again gets into the circulation of the blood and makes the previously mentioned disorders still greater. And if these bodily affections be severe, still worse are those which precede them; as when the bone itself, by reason of the density of the flesh, does not receive sufficient air, but becomes stagnant and hot and gangrened and receives no nutriment, and the natural process is inverted, and the bone crumbling passes into the food, and the food into the flesh, and the flesh again falling into the blood causes maladies yet more violent than those already mentioned. But the worst of all is when the marrow is diseased, either from excess or defect; and this is the cause of the very greatest and most fatal disorders in which the whole course of the body is reversed. There is a third class of diseases which may be conceived of as arising in three ways, and are produced sometimes by wind, and sometimes by phlegm, and sometimes by bile. When the lung, which is to the body the steward of the air, is obstructed by rheums and has the passages stopped up, having no egress in one part, while in another part too much air enters in, then the parts which are unrefreshed by the air corrode, while in other parts the excess of wind forcing its way through the veins distorts them and consumes the body at the center, and is there shut in and holds fast the midriff; thus numberless painful diseases are produced, accompanied by copious sweats. And oftentimes when the flesh is dissolved in the interior of the body, wind, generated within and unable to get out, is the source of quite as much pain as the air coming in from without; but the greatest pain is when the wind gets about the sinews and the veins connected with them, and swells them up, especially when the pressure is upon the great sinews of the shoulder and twists back the ligaments that fasten them. These from the intensive nature of the affection, are termed tetanus and recurvation. The cure of them is difficult, and they generally end in fevers. The white phlegm, though dangerous when detained

85

within by reason of the air bubbles, yet being capable of relief by expiration, is less severe, and only discolors the body, generating white leprosies and similar diseases. When the phlegm is mingled with black bile and dispersed about the courses of the head, which are the divinest part of us, and disturbs them in sleep, the attack is not so severe; but when assailing those who are awake it is hard to be got rid of, and, being an affection of a sacred part, is most justly called sacred. An acid and salt phlegm, again, is the source of all those diseases which are of a catarrhal nature, but, because the places into which they flow are of various kinds, they have all sorts of names.

Inflammations of the body come from burnings and inflamings, and all of them originate in bile. When bile finds a means of discharge, it boils up and sends forth all sorts of tumors; but when kept down within, it generates many inflammatory diseases, above all when mingled with pure blood; as it then disturbs the order of the fibers which are scattered about in the blood and are designed to maintain the balance of rare and dense, in order that the blood may not by reason of heat perspire through the pores of the body, nor again become too dense and thus find a difficulty in circulating through the veins. The just temperament of these things is preserved by the fibers according to the appointment of nature; and if anyone collects them together when the blood is dead and congealed, then the blood that remains in them flows out, and thus left to themselves they also soon congeal with the surrounding cold. Such is the power which the fibers have of acting upon the blood; and from them arises bile, which is only stale blood, and from being flesh is liquefied again, and at the first influx comes in little by little warm and moist, and is congealed by the power of the fibers; and if congealed and extinguished by force produces internal cold and shuddering. But when it enters with more of a flood and overcomes the fibers by its heat, and makes them boil and bubble in a disorderly manner, if it have power enough completely to get the better, it passes into the marrow and burns up and unmoors what may be termed the cables of the ship, and frees the soul; but when there is not so much, and the body though wasted still holds out, it is either mastered and banished from the whole body, or is thrust through the veins into the lower or upper belly, and is driven out of the body like an exile out of an insurgent State, and causes diarrhœas and dysenteries, and all sorts of similar disorders. When the constitution is disordered by excess of fire, then the heat and fever are constant; when air is the cause, then the fever is quotidian; when water, which is a more sluggish element than either fire or air, then the fever intermits a day; when earth is the cause, which is the most sluggish of the four, and is only purged away in a fourfold period, the result is a quartan fever, which can only with difficulty be shaken off.

86

Such is the course of the diseases of the body; and the disorders of the soul which originate in the body are as follows: The disorder of the mind will be acknowledged to be folly; but there are two kinds of folly—one, madness, and the other ignorance; and whatever affection gives rise to either of them may be called disease. Excessive pains and pleasures are justly to be regarded as the greatest diseases of the soul, for a man who is in great joy or in great pain, in his irrational eagerness to attain the one and to avoid the other, is not truly able to see or to hear anything; but he is mad, and is at the same time quite incapable of any participation in reason. For he who has the seed about the spinal marrow too fruitful and prolific, like a tree overladen with fruit, has many throes, and also obtains many pleasures in his desires and their gratifications, and is for the most part of his life mad, because his pleasures and pains are so very great; his soul is rendered foolish and disorded by his body; and he is regarded not as one diseased, but as one who is voluntarily bad, which is a bad mistake. For the truth is that the intemperance of love for the most part grows into a disease of the soul by reason of the moist and fluid state of one element, and this arises out of the loose consistency of the bones. And in general, all that which is termed the intemperance of pleasure is unjustly charged upon those who do wrong, as if they did wrong voluntarily. For no man is voluntarily bad; but the bad become bad by reason of an ill disposition of the body and bad education: every man finds these things to be an evil and a mischief; and in like manner the soul is often vicious through the bodily influences of pain. For where the sharp and briny phlegm and other bitter and bilious humors wander over the body, and find no exit or escape, but are compressed within and mingle their own vapors with the motions of the soul, and are blended with them, they produce an infinite variety of diseases in all sorts of degrees, and being carried to the three places of the soul on which any of them may severally chance to alight, they create infinite varieties of trouble and melancholy, of tempers rash and cowardly, and also of forgetfulness and stupidity. Further, when men's bodies are thus ill made, and evil forms of government are superadded—when in States evil discourses are uttered in private as well as in public, and when from youth upward no sort of instruction is given which may heal these ills, here is another source of evil; and these are the two ways in which all of us who are bad become bad, through two things which are wholly out of our power. And for this the planters are to blame rather than the plants, the educators rather than the educated. Still we should endeavor as far as we can by education, and studies, and learning, to avoid vice and attain virtue; this, however, is part of another subject.

There is a corresponding inquiry concerning the modes in which the mind and the body are to be treated, and by what means they are preserved, on which I may and ought to enter; for it is more our duty to speak of the good than of the evil. Everything that is good is fair, and the fair is not without measure, and the animal who is fair may be supposed to have measure.

Now we perceive lesser symmetries and comprehend them, but about the highest and greatest we have no understanding; for with a view to health and disease, and virtue and vice, there is no symmetry or want of symmetry greater than that of the soul to the body; and this we do not perceive, or ever reflect that when a weaker or lesser frame is the vehicle of a great and mighty soul, or conversely, when they are united in the opposite way, then the whole animal is not fair, for it is defective in the most important of all symmetries; but the fair mind in the fair body will be the fairest and loveliest of all sights to him who has the seeing eye. Just as a body which has a leg too long, or some other disproportion, is an unpleasant sight, and also, when undergoing toil, has many sufferings, and makes violent efforts, and often stumbles through awkwardness, and is the cause of infinite evil to its own self—in like manner we should conceive of the double nature which we call the living being; and when in this compound there is an impassioned soul more powerful than the body, that soul, I say, convulses and disorders the whole inner nature of man; and when too eager in the pursuit of knowledge, causes wasting; or again, when teaching or disputing in private or in public, and strifes and controversies arise, inflames and dissolves the composite frame of man and introduces rheums; and the nature of this is not understood by most professors of medicine, who ascribe the phenomenon to the opposite of the real cause. And once more, when a body large and too much for the soul is united to a small and weak intelligence, seeing that there are two desires natural to man, one of food for the sake of the body, and one of wisdom for the sake of the diviner part of us—then, I say, the motions of the stronger principal, getting the better and increasing their own power, but making the soul dull, and stupid, and forgetful, engender ignorance, which is the greatest of diseases. There is one protection against both, that we should not move the body without the soul or the soul without the body, and thus they will aid one another, and be healthy and well balanced. And

88

therefore the mathematician or anyone else who devotes himself to some intellectual pursuit, must allow his body to have motion also, and practice gymnastic; and he who would train the limbs of the body, should impart to them the motions of the soul, and should practice music and all philosophy, if he would be called truly fair and truly good. And in like manner should the parts be treated, and the principle of the whole similarly applied to them; for as the body is heated and also cooled within by the elements which enter in, and is again dried up and moistened by external things, and experiences these and the like affections from both kinds of motions, the result is that the body if given up to motion when in a state of quiescence is overmastered and destroyed; but if anyone, in imitation of that which we call the foster-mother and nurse of the universe, will not allow the body to be at rest, but is always producing motions and shakings, which constantly react upon the natural motions both within and without, and by shaking moderately the affections and parts which wander about the body, brings them into order and affinity with one another according to the theory of the universe which we were maintaining, he will not allow enemy placed by the side of enemy to create wars and disorders in the body, but he will place friend by the side of friend, producing health. Now of all motions that is the best which is produced in a thing by itself, for it is most akin to the motion of the intelligent and the motion of the universe; but that motion which is caused by others is not so good, and worst of all is that which moves the parts of the body, when prostrate and at rest, in parts only and by external means; wherefore also that is the best of the purifications and adjustments of the body which is affected by gymnastic; next is that which is effected by carrying the body, as in sailing or any other mode of conveyance which is not fatiguing; the third sort of motion may be of use in a case of extreme necessity, but in any other will be adopted by no man of sense: I mean the purgative treatment of physicians;

89

for diseases which are not attended by great dangers should not be irritated by purgatives, for every form of disease is in a manner akin to the living being—for the combination out of which they were formed has an appointed term of life and of existence. And the whole race and every animal has his appointed natural time, apart from violent casualties; for the triangles are originally framed with power to live for a certain time, beyond which no man can prolong his life. And this holds also of the nature of diseases, for if anyone regardless of their appointed time would destroy nature by purgatives, he only increases and multiplies them. Wherefore we ought always to manage them by regimen, as far as a man can spare the time, and not provoke a disagreeable enemy by medical treatment. Let this much be said of the general nature of man, and of the body which is a part of him, and of the manner in which a man may govern himself and be governed best, and live most according to reason: and we must begin by providing that the governing principle shall be the fairest and best possible for the purpose of government. But to discuss such a subject accurately would be a sufficiently long business of itself. As a mere supplement or sequel of what has preceded, it may be summed up as follows: As I have often said, that there are three kinds of soul located within us, each of them having their own proper motions—so I must now say in the fewest words possible, that the one part, if remaining inactive and ceasing from the natural motion, must necessarily become very weak, but when trained and exercised then very strong.

90 Wherefore we should take care that the three parts of the soul are exercised in proportion to one another.

Concerning the highest part of the human soul, we should consider that God gave this as a genius to each one, which was to dwell at the extremity of the body, and to raise us like plants, not of an earthly but of a heavenly growth, from earth to our kindred which is in heaven. And this is most true; for the divine power suspended the head and root of us from that place where the generation of the soul first began, and thus made erect the whole body. He, therefore, who is always occupied with the cravings of desire and ambition, and is eagerly striving after them, must have all his opinions mortal, and, as far as man can be, must be all of him mortal, because he has cherished his mortal part. But he who has been earnest in the love of knowledge and true wisdom, and has been trained to think that these are the immortal and divine things of a man, if he attain truth, must of necessity, as far as human nature is capable of attaining immortality, be all immortal, as he is ever serving the divine power; and having the genius residing in him in the most perfect order, he must be preëminently happy. Now there is only one way in which one being can serve another, and this is by giving him his proper nourishment and motion. And the motions which are akin to the divine principle within us are the thoughts and revolutions of the universe. These each man should follow, and correct those corrupted courses of the head which are concerned with generation, and by learning the harmonies and revolutions of the whole, should assimilate the perceiver to the thing perceived, according to his original nature, and by thus assimilating them, attain that final perfection of life, which the gods set before mankind as best, both for the present and the future.

Thus the discussion of the universe which according to our original proposition, was to reach to the origin of man, seems to have an end. A brief mention may be made of the generation of other animals, but there is no need to dwell upon them at length; this would seem to be the best mode of attaining a due proportion. On the subject of animals, then, the following remarks may be offered. Of the men who came into the world, those who are cowards or have led unjust lives may be fairly supposed to change into the nature of women in the second generation. Wherefore also at the time when this took place the gods created in us the desire of generation,
91

contriving in man one animated substance, and in woman another, which they formed respectively in the following manner: The passage for the drink by which liquids pass through the lung under the kidneys and into the bladder, and which receives and emits them by the pressure of the breath, was so fashioned as to penetrate also into the body of the marrow, which passes from the head along the neck and through the back, and which in our previous discussion we have named the seed. And the seed having life, and becoming endowed with respiration, produces, in that part in which it respires, a lively desire of emission, and thus creates in us the love of procreation. Wherefore also in men the organ of generation becoming rebellious and masterful, like an animal disobedient to reason, seeks, by the raging of the appetites, to gain absolute sway; and the same is the case with the wombs and other organs of women; the animal within them is desirous of procreating children, and when remaining without fruit long beyond its proper time, gets discontented and angry, and wandering in every direction through the body, closes up the passages of the breath, and, by obstructing respiration, drives them into the utmost difficulty, causing all varieties of disease, until at length the desire and love of the man and the woman, as it were producing and plucking the fruit from the tree, cause the emission of seed into the womb, as into a field, in which they sow animals unseen by reason of their smallness, and formless; these they again separate and mature them within, and after that bring them out into the light, and thus perfect the generation of animals.

Thus were created women and the female sex in general. But the race of birds was created out of innocent light-minded men who, although their thoughts were directed towards heaven, imagined, in their simplicity, that the clearest demonstration of the things above was to be obtained by sight; these were transformed into birds, and they grew feathers instead of hair.

The race of wild pedestrian animals again came from those who had no philosophy in all their thoughts, and never considered at all about the nature of the heavens, because they had ceased to use the courses of the head, and followed the guidance of those parts of the soul which surround the breast. In consequence of these habits of theirs they had their forelegs and heads trailing upon the earth to which they were akin; and they had also the crowns of their heads oblong, and in all sorts of curious shapes, in which the courses of the soul were compressed by reason of disuse. And this was the reason why quadrupeds and polypods were created: God gave the more senseless of them the more support that they might be more attracted to the earth. And the most foolish of them, who trailed their bodies entirely upon the ground and have no longer any need of feet, he made without feet to crawl upon the earth. The fourth class were the inhabitants of the water: these were made out of the most entirely ignorant and senseless beings, whom the transformers did not think any longer worthy of pure respiration, because they possessed a soul which was made impure by all sorts of transgression; and instead of allowing them to respire to the subtle and pure element of air, they thrust them into the water and gave them a deep and muddy medium of respiration; and hence arose the race of fishes and oysters, and other aquatic animals, which have received the most remote habitations as a punishment of their extreme ignorance. These are the laws by which animals pass into one another, both now and ever changing as they lose or gain wisdom and folly.

92

And now we may say that our discourse about the nature of the universe has come to an end. The world has received animals, mortal and immortal, and is fulfilled with them, and has become a visible animal comprehending the visible, the sensible God who is the image of the intelligible, greatest, best, fairest, and most perfect, the one only-begotten universe.

CRITIAS

Persons of the Dialogue

CRITIAS.   TIMAEUS.

HERMOCRATES.   SOCRATES.

TIM. HOW THANKFUL I AM, SOCRATES, THAT I HAVE ARRIVED AT last, and, like a weary traveler after a long journey, may now be at rest! And I pray the being who always was of old, and has now been by me declared, to receive and preserve my words, in so far as they have been spoken truly and acceptably to him; and if unintentionally I have said anything wrong, I pray that he will impose upon me a fitting retribution, and the proper retribution of him who errs is to set him in the right way. Wishing, then, that for the future I may speak truly concerning the generation of the gods, I pray them to give me knowledge, which of all medicines is the most perfect and best. That is my prayer. And now I deliver the argument into the hands of Critias, according to our agreement.

CRIT. And I, Timaeus, accept the trust, and as you at first said that you were going to speak of high matters, and begged that some allowance might be extended to you, I must request the same or a greater allowance for what I am about to say. And 107 although I very well know that I am making an ambitious and a somewhat rude request, I must not be deterred by that. For will any man of sense deny that you have spoken well? I can only attempt to show that my theme is more difficult, and claims more indulgence than yours; and I shall argue that to seem to speak well of the gods to men is far easier than to speak well of mortals to one another: for the inexperience and utter ignorance of his hearers about such matters is a great assistance to him who has to speak of them, and we know how ignorant we are concerning the gods. But I should like to make my meaning clearer, if you will follow me. All that we are any of us saying can only be imitation and assimilation. For if we consider how the works of the painter represent bodies divine and heavenly, and the different degrees of gratification with which the eye of the spectator receives them, we shall see that we are satisfied with the artist who is able in any degree to imitate the earth and its mountains, and the rivers, and the woods, and the universe, and the things that are and move therein, and further, that knowing nothing precise about such matters, we do not examine or analyze the painting; all that is required is a sort of indistinct and deceptive mode of shadowing them forth. But when a person endeavors to paint the human form we are quick at finding out defects, and our familiar knowledge makes us severe judges of anyone who does not render every point of similarity; and this is also true of discourse; we are satisfied with a picture of divine and heavenly things which has very little likeness to them; but we are more precise in our criticism of mortal and human things. Wherefore if at the moment of speaking we cannot suitably express what we mean, you must excuse us, considering that to form approved likenesses of human things is the reverse of easy. This is what I want to suggest to you, and at the same 108 time to beg, Socrates, that I may have not less, but more indulgence conceded to me in what I am about to say. Which favor, if I am right in asking, I hope that you will be ready to grant.

Soc. Certainly, Critias, we will grant that, and we will grant the same by anticipation to Hermocrates, who has to speak third; for I have no doubt that when his turn comes a little while hence, he will make the same request which you have made. In order, then, that he may provide himself with a fresh beginning, and not be compelled to say the same things over again, let him understand that the indulgence is already extended by anticipation to him. And now, friend Critias, I will announce to you the judgment of the theater. They are of opinion that the last performer was wonderfully successful, and that you will need a great deal of indulgence if you are to rival him.

HER. The warning, Socrates, which you have addressed to him, I must also regard as applying to myself. But remember, Critias, that faint heart never yet raised a trophy; you must go and attack the argument like a man. First invoke Apollo and the Muses, and then let us hear you sing the praises of your ancient citizens.

CRIT. Friend Hermocrates, you who are stationed last and have another in front of you, have not lost heart as yet; whether you are right or not, you will soon know; meanwhile I accept your exhortations and encouragements. But in addition to the gods whom you have mentioned, I would specially invoke Mnemosyne; for all the important part of what I have to tell is dependent on her favor, and if I can recollect and recite enough of what was said by the priests and brought hither by Solon, I doubt not that I shall satisfy the requirements of this theater. To that task then I will at once address myself.

Let me begin by observing first of all, that nine thousand was the sum of years which had elapsed since the war which was said to have taken place between all those who dwelt outside the pillars of Heracles and those who dwelt within them; this war I am now to describe. Of the combatants on the one side, the city of Athens was reported to have been the ruler and to have directed the contest; the combatants on the other side were led by the kings of the islands of Atlantis, which, as I was saying, once had an extent greater than that of Libya and Asia; and when afterwards sunk by an earthquake, became an impassable barrier of mud to voyagers sailing from hence to the ocean. The progress of the history will unfold the various tribes of

109

barbarians and Hellenes which then existed, as they successively appear on the scene; but I must begin by describing first of all the Athenians, as they were in that day, and their enemies who fought with them; and I shall have to tell of the power and form of government of both of them. Let us give the precedence to Athens:

In former ages, the gods had the whole earth distributed among them by allotment; there was no quarreling; and you cannot suppose that the gods did not know what was proper for each of them to have; or, knowing this, that they would seek to procure for themselves by contention that which more properly belonged to others. Each of them obtained righteously by lot what they wanted, and peopled their own districts; and when they had peopled them they tended us human beings who belonged to them as shepherds tend their flocks, excepting only that they did not use blows or bodily force, as the manner of shepherds is, but governed us like pilots from the stern of a vessel, which is an easy way of guiding animals, by the rudder of persuasion, taking hold of our souls according to their own pleasure; thus did they guide all mortal creatures. Now different gods had their inheritance in different places which they set in order. Hephaestus and Athene, who were brother and sister, and sprang from the same father, having a common nature, and being united also in the love of philosophy and of art, both obtained as their allotted region this land, which was naturally adapted for wisdom and virtue; and there they implanted brave children of the soil, and put into their minds the order of government; their names are preserved, but their actions have disappeared by reason of the destruction of those who had the tradition, and the lapse of ages. For the survivors of each destruction, as I have already said, dwelt in the mountains; they were ignorant of the art of writing, and had heard only the names of the chiefs of the land, and a very little about their actions. The names they gave to their children out of affection, but of the virtues and laws of those who preceded them, they knew only by obscure traditions; and as they themselves and their children were for many generations in want of the necessaries of life, they directed their attention to the supply of their wants, and of that they discoursed, to the neglect of events that had happened in times long passed; my mythology and the inquiry into antiquity are introduced into cities when they have leisure, and when they see the necessaries of life already beginning to be provided, but not before. And this is the reason why the names of the ancients have been preserved to us without their deeds. This I infer because Solon said that the priests in their narrative of that war mentioned most of the names which are recorded prior to the time of Theseus, such as Cecrops, and Erechtheus, and Erichthonius, and Erysichthon, and the names of the women in like manner. Moreover, the figure and image of the goddess show that at that time military pursuits were common to men and women, and that in accordance with that custom they dedicated the armed image of the goddess as a testimony that all animals, male and female, which consort together, have a virtue proper to each class, which they are all able to pursue in common.

110

Now the country was inhabited in those days by various classes of citizens: there were artisans, and there were husbandmen, and there was a warrior class originally set apart by divine men; these dwelt by themselves, and had all things suitable for nurture and education; neither had any of them anything of their own, but they regarded all things as common property; nor did they require to receive of the other citizens anything more than their necessary food. And they practiced all the pursuits which we yesterday described as those of our imaginary guardians. Also about the country the Egyptian priests said what is not only probable but also true, that the boundaries were fixed by the Isthmus, and that in the other direction they extended as far as the heights of Cithaeron and Parnes; the boundary line came down towards the plain, having the district of Oropus on the right, and the river Asopus on the left, as the limit towards the sea. The land was the best in the world, and for this reason was able in those days to support a vast army, raised from the surrounding people. And a great proof of this fertility is, that the part which still remains may compare with any in the world for the variety and excellence of its fruits and the suitableness of its pastures to every sort of animal; and besides beauty the land had also plenty. How am I to prove this? And of what remnant of the land then in existence may this be truly said? I would have you observe the present aspect of the country, which is only a promontory extending far into the sea away from the rest of the continent, and the surrounding basin of the sea is everywhere deep in the neighborhood of the shore. Many great deluges have taken place during the nine thousand years, for that is the number of years which have elapsed since the time of which I am speaking; and in all the ages and changes of things, there has never been any settlement of the earth flowing down from the mountains as in other places, which is worth speaking of; it has always been carried round in a circle and disappeared in the depths below. The consequence is, that in comparison of what then was, there are remaining in small islets only the bones of the wasted body, as they may be called; all the richer and softer parts of the soil having fallen away, and the mere skeleton of the country being left. But in former days, and in the primitive state of the country, what are now mountains were only regarded as hills; and the plains, as they are now termed, of Phelleus were full of rich earth, and there was abundance of wood in the mountains. Of this last the traces still remain, for there are some of the mountains which now only afford sustenance to bees, whereas not long ago there were still remaining roofs cut from the trees growing there, which were of a size sufficient to cover the largest houses; and there were many other high trees, bearing fruit, and abundance of food for

111

cattle. Moreover, the land enjoyed rain from heaven year by year, not, as now, losing the water which flows off the earth into the sea, but having an abundance in all places, and receiving and treasuring up in the close clay soil the water which drained from the heights, and letting this off into the hollows, providing everywhere abundant streams of fountains and rivers; and there may still be observed indications of them in ancient sacred places, where there are fountains; and this proves the truth of what I am saying.

Such was the natural state of the country, which was cultivated, as we may well believe, by true husbandmen, who were lovers of honor, and of a noble nature, and did the work of husbandmen, and had a soil the best in the world, and abundance of water, and in the heaven above an excellently tempered climate.

112 Now the city in those days was arranged on this wise; in the first place the Acropolis was not as now. For the fact is that a single night of excessive rain washed away the earth and laid bare the rock; at the same time there were earthquakes, and then occurred the third extraordinary inundation, which immediately preceded the great destruction of Deucalion. But in primitive times the hill of the Acropolis extended to the Eridanus and Ilissus, and included the Pnyx and the Lycabettus as a boundary on the opposite side to the Pnyx, and was all well covered with soil, and level at the top, except in one or two places. Outside the Acropolis and on the sides of the hill there dwelt artisans, and such of the husbandmen as were tilling the ground near; at the summit the warrior class dwelt by themselves around the temples of Athene and Hephaestus, living as in the garden of one house, and surrounded by one inclosure. On the north side they had common houses, and had prepared for themselves winter places for common meals, and had all the buildings which they needed for the public use, and also temples, but unadorned with gold and silver, for these were not in use among them; they took a middle course between meanness

and extravagance, and built moderate houses in which they and their children's children grew old, and handed them down to others who were like themselves, always the same. And in summer time they gave up their gardens and gymnasia and common tables and used the southern quarter of the Acropolis for such purposes. Where the Acropolis now is there was a single fountain, which was extinguished by the earthquake, and has left only a few small streams which still exist, but in those days the fountain gave an abundant supply of water, which was of equal temperature in summer and winter. This was the fashion in which they lived, being the guardians of their own citizens and the leaders of the Hellenes, who were their willing followers. And they took care to preserve the same number of men and women for military service, which was to continue through all time, and still is, that is to say, about twenty thousand. Such were the ancient Athenians, and after this manner they righteously administered their own land and the rest of Hellas; they were renowned all over Europe and Asia for the beauty of their persons and for the many virtues of their souls, and were more famous than any of their contemporaries. And next, if I have not forgotten what I heard when I was a child, I will impart to you the character and origin of their adversaries. For friends should not keep their stories to themselves, but have them in common.

Yet, before proceeding further in the narrative, I ought to warn you that you must not be surprised if you should hear Hellenic names given to foreigners. I will tell you the reason of this: Solon, who was intending to use the tale for his poem, made an investigation into the meaning of the names, and found that the early Egyptians in writing them down had translated them into their own language, and he recovered the meaning of the several names and re-translated them, and copied them out again in our language. My great-grandfather, Dropidas, had the original writing, which is still in my possession, and was carefully studied by me when I was a child. Therefore

113

if you hear names such as are used in this country, you must not be surprised, for I have told you the reason of them. The tale, which was of great length, began as follows:

I have before remarked in speaking of the allotments of the gods, that they distributed the whole earth into portions differing in extent, and made themselves temples and sacrifices. And Poseidon, receiving for his lot the island of Atlantis, begat children by a mortal woman, and settled them in a part of the island, which I will proceed to describe. On the side towards the sea and in the center of the whole island, there was a plain which is said to have been the fairest of all plains and very fertile. Near the plain again, and also in the center of the island at a distance of about fifty stadia, there was a mountain not very high on any side. In this mountain there dwelt one of the earth-born primeval men of that country, whose name was Evenor, and he had a wife named Leucippe, and they had an only daughter who was named Cleito. The maiden was growing up to womanhood, when her father and mother died; Poseidon fell in love with her and had intercourse with her, and breaking the ground, inclosed the hill in which she dwelt all round, making alternate zones of sea and land larger and smaller, encircling one another; there were two of land and three of water, which he turned as with a lathe, out of the center of the island, equidistant every way, so that no man could get to the island, for ships and voyages were not as yet heard of. He himself, as he was a god, found no difficulty in making special arrangements for the center island, bringing two streams of water under the earth, which he caused to ascend as springs, one of warm water and the other of cold, and making every variety of food to spring up abundantly in the earth. He also begat and brought up five pairs of male children, dividing the island of Atlantis into ten portions; he gave to the firstborn of the eldest pair his mother's dwelling and the surrounding allotment, which was the largest and best, and made him king over the rest; the others he made

114

princes, and gave them rule over many men, and a large territory. And he named them all; the eldest who was king, he named Atlas, and from him the whole island and the ocean received the name of Atlantic. To his twin brother, who was born after him, and obtained as his lot the extremity of the island towards the pillars of Heracles as far as the country which is still called the region of Gades in that part of the world, he gave the name which in the Hellenic language is Eumelus, in the language of the country which is named after him, Gadeirus. Of the second pair of twins he called one Ampheres, and the other Evaemon. To the third pair of twins he gave the name Mneseus to the elder, and Autochthon to the one who followed him. Of the fourth pair of twins he called the elder Elasippus, and the younger Mestor. And of the fifth pair he gave to the elder the name of Azaes, and to the younger that of Diaprepes. All these and their descendants were the inhabitants and rulers of divers islands in the open sea; and also, as has been already said, they held sway in the other direction over the country within the pillars as far as Egypt and Tyrrhenia. Now Atlas had a numerous and honorable family, and his eldest branch always retained the kingdom, which the eldest son handed on to his eldest for many generations; and they had such an amount of wealth as was never before possessed by kings and potentates, and is not likely ever to be again, and they were furnished with everything which they could have, both in city and country. For because of the greatness of their empire many things were brought to them from foreign countries, and the island itself provided much of what was required by them for the uses of life. In the first place, they dug out of the earth whatever was to be found there, mineral as well as metal, and that which is now only a name and was then something more than a name, orichalcum, was dug out of the earth in many parts of the island, and with the exception of gold was esteemed the most precious of metals among the men of those days. There was an abundance

of wood for carpenter's work, and sufficient maintenance for tame and wild animals. Moreover, there were a great number of elephants in the island, and there was provision for animals
115 of every kind, both for those which live in lakes and marshes and rivers, and also for those which live in mountains and on plains, and therefore for the animal which is the largest and most voracious of them. Also whatever fragrant things there are in the earth, whether roots, or herbage, or woods, or distilling drops of flowers or fruits, grew and thrived in that land; and again, the cultivated fruit of the earth, both the dry edible fruit and other species of food, which we call by the general name of legumes, and the fruits having a hard rind, affording drinks and meats and ointments, and good store of chestnuts and the like, which may be used to play with, and are fruits which spoil with keeping, and the pleasant kinds of dessert, which console us after dinner, when we are full and tired of eating—all these that sacred island lying beneath the sun, brought forth fair and wondrous in infinite abundance. All these things they received from the earth, and they employed themselves in constructing their temples and palaces and harbors and docks; and they arranged the whole country in the following manner:

First of all they bridged over the zones of sea which surrounded the ancient metropolis, and made a passage into and out of the royal palace; and then they began to build the palace in the habitation of the god and of their ancestors. This they continued to ornament in successive generations, every king surpassing the one who came before him to the utmost of his power, until they made the building a marvel to behold for size and for beauty. And beginning from the sea they dug a canal of three hundred feet in width and one hundred feet in depth, and fifty stadia in length, which they carried through to the outermost zone, making a passage from the sea up to this, which became a harbor, and leaving an opening sufficient to enable the largest vessels to find ingress. Moreover, they divided the zones of land which parted the zones of sea, constructing bridges of such a width as would leave a passage for a single trireme to pass out of one into another, and roofed them over; and there was a way underneath for the ships; for the banks of the zones were raised considerably above the water. Now the largest of the zones into which a passage was cut from the sea was three stadia in breadth, and the zone of land which came next of equal breadth; but the next two, as well the zone of water as of land, were two stadia, and the one which surrounded the central island was a stadium only in width. The island in 116 which the palace was situated had a diameter of five stadia. This and the zones and the bridge, which was the sixth part of a stadium in width, they surrounded by a stone wall, on either side placing towers, and gates on the bridges where the sea passed in. The stone which was used in the work they quarried from underneath the center island, and from underneath the zones, on the outer as well as the inner side. One kind of stone was white, another black, and a third red, and as they quarried, they at the same time hollowed out docks double within, having roofs formed out of the native rock. Some of their buildings were simple, but in others they put together different stones which they intermingled for the sake of ornament, to be a natural source of delight. The entire circuit of the wall, which went round the outermost one, they covered with a coating of brass, and the circuit of the next wall they coated with tin, and the third, which encompassed the citadel, flashed with the red light of orichalcum. The palaces in the interior of the citadel were constructed on this wise: In the center was a holy temple dedicated to Cleito and Poseidon, which remained inaccessible, and was surrounded by an inclosure of gold; this was the spot in which they originally begat the race of the ten princes, and thither they annually brought the fruits of the earth in their season from all the ten portions, and performed sacrifices to each of them. Here, too, was Poseidon's own temple of a stadium in length, and half a stadium in width and of a proportionate height, having a sort of barbaric splendor. All the outside of the temple, with the exception of the pinnacles, they covered with silver, and the pinnacles with gold. In the interior of the temple the roof was of ivory, adorned everywhere with gold and silver and orichalcum; all the other parts of the walls and pillars and floor they lined with orichalcum. In the temple they placed statues of gold; there was the god himself standing in a chariot—the charioteer of six winged horses—and of such a size that he touched the roof of the buildings with his head; around him there were a hundred Nereids riding on dolphins, for such was thought to be the number of them in that day. There were also in the interior of the temple other images which had been dedicated by private individuals. And around the temple on the outside were placed statues of gold of all the ten kings and of their wives, and there were many other great offerings both of kings and of private individuals, coming both from the city itself and the foreign cities over which they held sway. There was an altar too, which in size and workmanship corresponded to the rest of the work, and there were palaces, in like manner, which answered to the greatness of the kingdom, and the glory of the temple.

117

In the next place, they used fountains both of cold and hot springs; these were very abundant, and both kinds[1] wonderfully adapted to use by reason of the sweetness and excellence of their waters. They constructed buildings about them and planted suitable trees; also cisterns, some open to the heaven, others which they roofed over, to be used in winter as warm baths; there were the king's baths, and the baths of private persons, which were kept apart; also separate baths for women, and others again for horses and cattle, and to each of them they gave as much adornment as was suitable for them. The water which ran off they carried, some to the grove of Poseidon, where were growing all manner of trees of wonderful height and beauty, owing to the excellence of the soil; the remainder was conveyed by aqueducts which passed over the bridges to the outer circles; and there were many temples built and dedicated to many gods; also gardens and places of exercise, some for men, and some set apart for horses, in both of the two islands formed by the zones; and in the center of the larger of the two there was a racecourse of a stadium in width, and in length allowed to extend all round the island, for horses to race in. Also there were guardhouses at intervals for the bodyguard, the more trusted of whom had their duties appointed to them in the lesser zone, which was nearer the Acropolis; while the most trusted of all had houses given them within the citadel, and about the persons of the kings. The docks were full of triremes and naval stores, and all things were quite ready for use. Enough of the plan of the royal palace. Crossing the outer harbors, which were three in number, you would come to a wall which began at the sea and went all round: this was everywhere distant fifty stadia from the largest zone and harbor, and inclosed the whole, meeting at the mouth of the channel towards the sea. The entire area was densely crowded with habitations; and the canal and the largest of the harbors were full of vessels and merchants coming from all parts, who, from their numbers, kept up a multitudinous sound of human voices and din of all sorts night and day.

I have repeated his descriptions of the city and the parts about the ancient palace nearly as he gave them, and now I must endeavor to describe the nature and arrangement of the rest of the country. The whole country was described 118 as being very lofty and precipitous on the side of the sea, but the country immediately about and surrounding the city was a level plain, itself surrounded by mountains which descended towards the sea; it was smooth and even, but of an oblong shape, extending in one direction three thousand stadia, and going up the country from the sea, through the center of the island, two thousand stadia; the whole region of the island lies towards the south, and is sheltered from the north. The surrounding mountains he celebrated for their number and size and beauty, in which they exceeded all that are now to be seen anywhere; having in them also many wealthy inhabited villages, and rivers, and lakes, and meadows supplying food enough for every animal, wild or tame, and wood of various sorts, abundant for every kind of work.

I will now describe the plain, which had been cultivated during many ages by many generations of kings. It was rectangular, and for the most part straight and oblong; and what it wanted of the straight line followed the line of the circular ditch. The depth, and width, and length of this ditch were incredible, and gave the impression that such a work, in addition to so many other works, could hardly have been wrought by the hand of man. But I must say what I have heard. It was excavated to the depth of a hundred feet, and its breadth was a stadium everywhere; it was carried round the whole of the plain, and was ten thousand stadia in length. It received the streams which came down from the mountains, and winding round the plain and touching the city at various points, was there let off into the sea. From above, likewise, straight canals of a hundred feet in width were cut in the plain, and again let off into the ditch towards the sea: these canals were at intervals of an hundred stadia, and by them they brought down the wood from the mountains to the city, and conveyed the fruits of the earth in ships, cutting transverse passages from one canal into another, and to the city. Twice in the year they gathered the fruits of the earth—in winter having the benefit of the rains, and in summer introducing the water of the canals.

As to the population, each of the lots in the plain had an appointed chief of men who were fit for military service, and the size of the lot was to be a square of ten stadia each way, and the total number of all the lots was sixty thousand. And

119

of the inhabitants of the mountains and of the rest of the country there was also a vast multitude having leaders, to whom they were assigned according to their dwellings and villages. The leader was required to furnish for the war the sixth portion of a war-chariot, so as to make up a total of ten thousand chariots; also two horses and riders upon them, and a light chariot without a seat, accompanied by a fighting man on foot carrying a small shield, and having a charioteer mounted to guide the horses; also, he was bound to furnish two heavy armed, two archers, two slingers, three stone-shooters, and three javelin-men, who were skirmishers, and four sailors to make up a complement of twelve hundred ships. Such was the order of war in the royal city—that of the other nine governments was different in each of them, and would be wearisome to narrate.

As to offices and honors, the following was the arrangement from the first. Each of the ten kings in his own division and in his own city had the absolute control of the citizens, and in many cases, of the laws, punishing and slaying whomsoever he would. Now the relations of their governments to one another were regulated by the injunctions of Poseidon as the law had handed them down. These were inscribed by the first men on a column of orichalcum, which was situated in the middle of the island, at the temple of Poseidon, whither the people were gathered together every fifth and sixth years alternately, thus giving equal honor to the odd and to the even number. And when they were gathered together they consulted about public affairs, and inquired if anyone had transgressed in anything, and passed judgment on him accordingly, and before they passed judgment they gave their pledges to one another on this wise: There were bulls who had the range of the temple of Poseidon; and the ten who were left alone in the temple, after they had offered prayers to the gods that they might take the sacrifices which were acceptable to them, hunted the bulls, without weapons, but with staves and nooses; and the bull

which they caught they led up to the column; the victim was
then struck on the head by them and slain over the sacred inscrip-
tion. Now on the column, besides the law, there was inscribed
an oath invoking mighty curses on the disobedient. When
therefore, after offering sacrifice according to their customs,
they had burnt the limbs of the bull, they mingled a cup and
cast in a clot of blood for each of them; the rest of the victim
they took to the fire, after having made a purification of the
column all round. Then they drew from the cup in golden
vessels, and pouring a libation on the fire, they swore that they
would judge according to the laws on the column, and would
punish anyone who had previously transgressed, and that
for the future they would not, if they could help, transgress
any of the inscriptions, and would not command or obey any
ruler who commanded them, to act otherwise than according
to the laws of their father Poseidon. This was the prayer which
each of them offered up for himself and for his family, at the
same time drinking and dedicating the vessel in the temple
of the god, and after spending some necessary time at supper,
when darkness came on and the fire about the sacrifice was
cool, all of them put on most beautiful azure robes, and, sitting
on the ground, at night, near the embers of the sacrifices on
which they had sworn, and extinguishing all the fire about
the temple, they received and gave judgment, if any of them
had any accusation to bring against anyone; and when they had
given judgment, at daybreak they wrote down their sentences
on a golden tablet, and deposited them as memorials with
their robes.

There were many special laws which the several kings
had inscribed about the temples, but the most important
was the following: That they were not to take up arms against
one another, and they were all to come to the rescue if anyone
in any city attempted to overthrow the royal house; like their
ancestors, they were to deliberate in common about war and

120

other matters, giving the supremacy to the family of Atlas. And the king was not to have the power of life and death over any of his kinsmen unless he had the assent of the majority of the ten kings.

Such was the vast power which the god settled in the lost island of Atlantis; and this he afterwards directed against our land on the following pretext, as traditions tell: For many generations, as long as the divine nature lasted in them, they were obedient to the laws, and well-affectioned towards the gods, who were their kinsmen; for they possessed true and in every way great spirits, practicing gentleness and wisdom in the various chances of life, and in their intercourse with one another. They despised everything but virtue, not caring for their present state of life, and thinking lightly of the possession of gold and other property, which seemed only a burden to them; neither were they intoxicated by luxury; nor did wealth deprive them of their self-control; but they were sober, and saw clearly that all these goods are increased by virtuous friendship with one another, and that by excessive zeal for them, and honor of them, the good of them is lost and friendship perishes with them. By such reflections and by the continuance in them of a divine nature, all that which we have described waxed and increased in them; but when this divine portion began to fade away in them, and became diluted too often and with too much of the mortal admixture, and the human nature got the upper hand, then they, being unable to bear their fortune, became unseemly, and to him who had an eye to see, they began to appear base, and had lost the fairest of their precious gifts; but to those who had no eye to see the true happiness, they still appeared glorious and blessed at the very time when they were filled with unrighteous avarice and power. Zeus, the god of gods, who rules with law, and is able to see into such things, perceiving that an honorable race was in a most wretched state, and wanting to inflict punishment on them, that they might be chastened and improve, collected all the gods into his most holy habitation, which being placed in the center of the world, sees all things that partake of generation. And when he had called them together, he spake as follows:

ENDNOTES

INTRODUCTION

[1] Owen, G. E. L "The Place of the *Timaeus* in Plato's Dialogues," *Classical Quarterly* (1953).

[2] Cherniss, H. F. "The Relation of the *Timaeus* to Plato's Later Dialogues," *Journal of Hellenic Studies* (1957).

[3] Keyt, D. "The Mad Craftsman of the *Timaeus*," *Philosophical Review* 80 (1971): 230–235.

[4] Vlastos, G. "The Disorderly Motion in the *Timaeus*," and "Creation in the *Timaeus*: Is It a Fiction?" in R. E. Allen, ed., *Studies in Plato's Metaphysics* (London: Routledge and Kegan Paul, 1965).

TIMAEUS

[1] Reading τὸ τῶν θηοευτῶν.

[2] E.g., 243 :: 256 : $\frac{81}{64}$: $\frac{4}{3}$ :: $\frac{243}{128}$: 2 :: $\frac{81}{32}$: $\frac{8}{3}$ :: $\frac{243}{64}$: 4 :: $\frac{81}{16}$: $\frac{16}{3}$ :: $\frac{342}{32}$: 8. (MARTIN).

[3] Reading αὐτό.

[4] He is speaking of two kinds of mirrors, first the plane, secondly the cylindrical and the latter is supposed to be placed, first vertically, secondly horizontally.

[5] Or taking μᾶλλον δὲ κ· τ· λ· with the preceding words, "but more probably"; or laying the stress on ἀπ᾽ ἀρχῆς, "but above and before all, I will begin at the beginning of each and all."

[6] Or reading ἅμμα, the knot.

[7] Reading χλοῶδες.

CRITIAS

[1] Reading ἑκατέρου πρὸς τὴν χρῆσιν.

105

SUGGESTED READING

MOHR, R. *The Platonic Cosmology.* Leiden: Brill, 1985.

PELIKAN, J. *What Has Athens to Do with Jerusalem? Timaeus and Genesis in Counterpoint.* Ann Arbor: University of Michigan Press, 1997.

PRIOR, W. J. *Unity and Development in Plato's Metaphysics.* La Salle, IL: Open Court, 1985.

ROBINSON, T. M. *Plato's Psychology.* Toronto: University of Toronto Press, 1970.

SORABJI, R. *Time, Creation and the Continuum.* Ithaca, NY: Cornell University Press, 1983.

VLASTOS, G. *Plato's Universe.* Seattle: University of Washington Press, 1975.

ZELLER, E. *Outlines of the History of Greek Philosophy.* 1883. Mineola, NY: Dover, 1980.